RESEARCH
METHODS

FOR CRIMINOLOGY
AND CRIMINAL JUSTICE

Third Edition

RESEARCH METHODS

FOR CRIMINOLOGY
AND CRIMINAL JUSTICE

Third Edition

M. L. DANTZKER, PhD
University of Texas, Pan American

RONALD D. HUNTER, PhD
Western Carolina University

JONES & BARTLETT
LEARNING

World Headquarters

Jones & Bartlett Learning
40 Tall Pine Drive
Sudbury, MA 01776
978-443-5000
info@jblearning.com
www.jblearning.com

Jones & Bartlett Learning
Canada
6339 Ormindale Way
Mississauga, Ontario L5V 1J2
Canada

Jones & Bartlett Learning
International
Barb House, Barb Mews
London W6 7PA
United Kingdom

Jones & Bartlett Learning books and products are available through most bookstores and online book-sellers. To contact Jones & Bartlett Learning directly, call 800-832-0034, fax 978-443-8000, or visit our website, www.jblearning.com.

Substantial discounts on bulk quantities of Jones & Bartlett Learning publications are available to corporations, professional associations, and other qualified organizations. For details and specific discount information, contact the special sales department at Jones & Bartlett Learning via the above contact information or send an email to specialsales@jblearning.com.

Production Credits
Publisher, Higher Education: Cathleen Sether
Acquisitions Editor: Sean Connelly
Associate Editor: Megan R. Turner
Associate Production Editor: Lisa Cerrone
Associate Marketing Manager: Lindsay White
Manufacturing and Inventory Control Supervisor: Amy Bacus
Composition: Paw Print Media
Cover Design: Kristin E. Parker
Photo and Permissions Associate: Emily O'Neill
Cover Image: © Tyler Olson/Dreamstime.com
Printing and Binding: Malloy, Inc.
Cover Printing: Malloy, Inc.

Library of Congress Cataloging-in-Publication Data Not Available at Time of Printing

ISBN: 978-0-7637-7732-6

6048

Printed in the United States of America
14 13 12 11 10   10 9 8 7 6 5 4 3 2 1

CONTENTS

The purpose of this text is to assist criminal justice and criminology students in developing an understanding (and hopefully an appreciation) of the basic principles of social science research. We do not seek to turn you into a research scientist in one short course of study, but we do hope that you will garner a better understanding of the research you may read. Furthermore, we will give you a rudimentary foundation that can be built upon, should you be interested in doing social science research, whether criminological or criminal justice oriented, in the future. This text will enable you to grasp the importance of scientific research, to read and comprehend all but the more complex research methodologies of others, and provide you with the basic tools to conduct your own social research.

Whether research is done by a college student completing a project for his or her degree (or just trying to understand an assigned reading) or by a professor meeting requirements or expectations associated with his or her position, it should be enjoyable and not a chore. The first step is to learn the basics for conducting research. A number of textbooks exist that can assist in this task, but many make learning about research—let alone conducting it—appear daunting. This text has made every attempt to ease the task of learning how to conduct research less daunting, and perhaps even to put the prospect of conducting research in a favorable light. To accomplish the above tasks this textbook is divided into three sections: Functions, Procedures, and Final Steps. Each chapter begins with a brief summary of what should be learned from the chapter. Within the text, realistic examples taken from currently published research are given to enhance the way the specific aspect of research is applicable to criminal justice and criminology.

Finally, each chapter ends with questions and/or exercises requiring students to apply what has been learned from the chapter.

The text begins by discussing what research is and why and how it is conducted. It addresses such questions as, What are criminal justice and criminological research? Why conduct this research? and, How can this research be completed? In general, it lays the foundation for conducting research.

Because criminal justice research often deals with human behavior, the ethics associated with such research is important. Chapter 2, Research and Ethics, discusses the ethics relevant to conducting research.

Deciding on what to conduct research can often be frustrating. However, there are numerous sources available to assist in making a decision on what to research. Chapter 3, The Beginning Basics, explores what sources to use and the issue of developing the research question, which is often the driving force behind social science research.

In Chapter 4, The Language of Research, students are introduced to researchese, or the terminology associated with conducting research, such as theory, hypothesis, population, sample, and variables. Furthermore, it briefly explores the processes required for conducting research through a researchese perspective.

Due to its long-standing image of being an applied social science and because of its lack of statistical sophistication, some of the research conducted and published has had its detractors. As a result, a debate continues as to what is more "academic," qualitative or quantitative research. Chapter 5, Qualitative Research, does not enter the debate but simply explains how this type of research fits into both criminology and criminal justice.

To help balance the debate over qualitative versus quantitative research, Chapter 6, Quantitative Research, takes over where Chapter 5 ends by exploring the other important types of research conducted in both criminology and criminal justice.

To successfully complete any type of research, it is important to establish a feasible plan or blueprint, known as the research design. Chapter 7, Research Designs, discusses the various research designs available for criminal justice and criminological research. They include historical, descriptive, inferential, developmental, case and field, correlational, and causal–comparative. A brief mention is made of true and quasi-experimental and action designs.

One of the most popular means of collecting data is the questionnaire. Although a rule of thumb is to use an established questionnaire, many individuals choose to design their own. Chapter 8, Questionnaire Construction, discusses the intricacies of designing a questionnaire, including issues of measurement, reliability, and validity.

It would be great if information could be gathered from a complete population, but this is almost impossible for criminal justice and criminological research. Therefore, sampling is an important aspect of research. In Chapter 9, Sampling, this concept and its related issues are examined.

In establishing the research design, a key component is how the data is to be collected. The four primary means for collecting data—survey, observation, archival, and unobtrusive means—are identified and explored within Chapter 10, Data Collection.

Once the data is collected, the question is what to do with it. There are a number of statistical techniques from which to choose. This is not a statistics book. However, to assist the student in better understanding the role of statistics in the research process we offer Chapters 11, Data Processing and Analysis, and Chapter 12, Inferential Statistics.

Now that the data is collected and analyzed, all there is left to do is write up the findings. For many, this is a daunting task. To help ease the fear and frustration, Chapter 13, Writing the Research, takes the student through a step-by-step introduction to writing the research.

Finally, to bring all the information offered throughout this text into a handy reference guide, we offer Chapter 14, Summing Up, an extensive, yet simple review of all the main concepts.

A final note concerns what this text is not. *Research Methods for Criminology and Criminal Justice, Third Edition*, is not a statistics book. However, it could be used in conjunction with a criminal justice statistics text. The fact is that separate books are often required to provide students the fullest extent of the knowledge required to conduct research and to analyze the data. This text allows students to learn how to conduct the research, leaving the statistics for another course and text.

We hope you will find the text as useful as it is intended to be. If nothing else, we hope it will help you at least feel more comfortable about reading or conducting criminal justice and criminological research.

Research Methods for Criminology and Criminal Justice is accompanied by PowerPoint chapter guides, lecture outlines, and a test bank for qualified adopters.

ACKNOWLEDGMENTS

As with each book, first edition or revisions, there are several people who deserve recognition and our gratitude. It all begins with the person who takes the chance of signing someone to write a book or resigning authors to revise an earlier edition. In this case it is Sean Connelly, Acquisitions Editor—our thanks for his support and confidence. Others deserving of recognition for their assistance are Megan Turner, Associate Editor; the copyeditor, Carla Jacobs; and Lisa Cerrone, Associate Production Editor. We would also like to offer our appreciation to the instructors who have used our book and made it possible to do a third edition.

We would also like to extend our thanks to the reviewers of the third edition manuscript:

Bruce B. Frey
University of Kansas

Evaristus Obinyan
Benedict College

Barbara Peat
Indiana University Northwest

Brenda Vose
University of North Florida

Finally, we'd like to thank our wives, Dr. Gail Dantzker and Mrs. Vi Hunter, for the constant love and support, especially during "crunch time." Thank you one and all.

Functions

Research:
What, Why, and How

What You Should Know!

Research methods for many graduate and undergraduate students can be misleading, confusing, and frustrating. The subject is often taught or presented in a manner where students may think that they are going to be social science researchers, actually conducting research. In reality, however, most students conduct little to no actual research but instead are consumers of research. That is, they read studies conducted by others perhaps hoping to apply to or better understand an area with which they may be working. Before being able to conduct research one should be able to understand the basics of research. Therefore, from this chapter the reader should be able to do the following:

1. Discuss tradition and authority as sources of human learning and be able to contrast their strengths and weaknesses.
2. Present and discuss the errors that plague casual observation.
3. Define what is meant by the scientific method. Explain how it seeks to remedy the errors of casual observation.
4. Compare and contrast the relationship between theory and research within the inductive and deductive logic processes.
5. Define research and explain its purpose.
6. Compare and contrast basic, applied, and multipurpose research.
7. Present and discuss the various types of research.
8. Present and discuss the reasons for criminologic research.

9. Present and discuss the various factors that influence research decisions.
10. Describe the primary steps in conducting research.

The Nature of Scientific Inquiry

It seems not that long ago that the authors were criminal justice students taking a first course in research methods. Our thoughts were, if we want to be police officers, why do we have to take this course? This is even worse than criminal theory, another useless course. What does it have to do with the real world in which we want to work? Later police experience in that real world taught us the value of both theory and research in the field of criminal justice. When we subsequently returned to school for graduate studies, the importance of theory and research was more readily apparent. We had learned that scientific investigation is very similar to criminal investigation: the use of a logical order and established procedures to solve real-world problems.

Social Science Research and the Real World

As police officers, the authors sought to determine whether a crime had been committed (what occurred and when it occurred); who had done it; how they had done it; and why they had done it. We then sought to use that investigatory knowledge to develop a successful prosecution of the offender. Our endeavors in the field taught us that the theory course that we had grudgingly endured had provided the rationale for human behavior on which the strategies of policing, courts, and corrections were based. We also discovered that those theories were not developed in some esoteric vacuum. They were the products of trial-and-error experiments conducted in policing, the courts, and corrections that had been refined and reapplied to their appropriate subject area. Today's police-deployment strategies, legal processes, and correctional techniques are all solidly based on prior theory and research.

These statements can also be applied to social science research in general. Typical real world conclusions are often flawed because of a number of issues that cause one's observations and reasoning to be inaccurate. The scientific method seeks to provide a means of investigation to correct (or a least limit) the inaccuracies of ordinary human inquiry (Adler & Clark, 2007; Bachman & Schutt, 2008; Kline, 2009). How one interprets their own observations and what one learns from others is based on tradition and authority. Tradition is the cultural teaching about the real world. "Poisonous snakes are dangerous. Beware of them!" You do not have to be bitten by a rattlesnake to appreciate its hazard. You have been taught by other members of your culture to respect the threat. This is an example of positive learning from tradition. It is based on the experiences of others in society who passed their knowledge on to others. Unfortunately, knowledge based on tradition is often erroneous. For example:

"Women are not suited to be police officers. They are too weak and too emotional." A multitude of highly competent and professional police officers have proved this sexist stereotype to be a fallacy.

The other source of secondhand knowledge is authority (Kraska & Neuman, 2008; Lavrakas, 2008; Maxfield & Babbie, 2009). Authority refers to new knowledge that is provided from the observations of others whom one respects. The cool aunt or uncle or older cousin who explained the facts of life to you was an authority figure. How accurate their explanations were, we leave to you to decide. As you got older you learned that much free advice was worth what you paid for it and that a great deal of bought advice also had little value. The importance of knowledge gained from authority figures depends on their qualifications relative to the subject being discussed. Therefore, one goes to a physician for help with health problems, and hires a plumber to fix a broken water pipe. These individuals are expected to have the expertise to provide solutions that laypersons do not have. Like tradition, the knowledge gained from dealings with authority figures can be extremely accurate or highly erroneous.

Science Versus Casual Inquiry

Casual inquiry is influenced by the sources of knowledge discussed in the previous section. In addition, there are other pitfalls that create errors in one's observations. It has been indicated that casual inquiry may be flawed because of inaccurate observation, overgeneralization, selective observation, and illogical reasoning (Kraska & Neuman, 2008; Lavrakas, 2008; Maxfield & Babbie, 2009).

Inaccurate observation occurs when conclusions are made based on hasty or incomplete observations. As an example, a young police officer once walked by a break room where a young records clerk was in tears. Sitting on each side of her were the captain in charge of internal affairs and an internal affairs investigator. The captain was telling her to stop crying in a harsh tone of voice. The officer immediately thought, "Those jerks. They could have at least taken her into their office before interrogating her." Several years later, the officer, then a sergeant for whom the woman in question now worked, learned that she had been extremely distraught over the break-up of her marriage and that the captain was a father figure to her who had actually been providing consolation.

Overgeneralization occurs when conclusions are made about individuals or groups based on knowledge of similar individuals or groups. "All lawyers are liars!" is an example. Despite the preponderance of lawyer jokes and any bad experiences one may have had with an attorney, one cannot accurately make that conclusion about all attorneys. There are simply too many attorneys (men and women of honesty and integrity and those of questionable ethics) to make such a conclusion without an individual knowledge of the person.

Selective observation is when one sees only those things that one wants to see. Racial and ethnic stereotyping is an example of negative selective observation. The attitude that "all whites are racists who seek to oppress" may cause the observer to see what he or she believes in the behaviors of all European Americans with whom they come into contact. Selective observation may also be positively biased. "My darling wonderful child has never done anything like that." Such selective observation can lead to major disappointment, such as when "He's a wonderful man who caters to my every whim" becomes "He's a selfish jerk who doesn't ever consider my feelings."

Finally, illogical reasoning happens when one decides that despite past observations, the future will be different. For example, the individual who plays the lottery week after week believing that eventually he has to win is an example of illogical reasoning. If the odds of success are unlikely, it is illogical to assume that by sheer willpower it can be made to occur.

Science seeks to reduce the possibility of the previously mentioned errors occurring by imposing order and rigor on observations. The means of doing so is the application of the scientific method.

The Scientific Method

The scientific method seeks to prevent errors of casual inquiry by using procedures that specify objectivity, logic, theoretical understanding, and knowledge of prior research in the development and use of a precise measurement instrument designed to record observations accurately (Bryman, 2008; Creswell, 2008; Gavin, 2008; McBurney & White, 2007). The result is a systematic search for the most accurate and complete description or explanation of the events or behaviors that are being studied. Just as a criminal investigation is a search for "the facts" and a criminal trial is a search for "the truth," the scientific method is a search for knowledge. The criminologic researcher seeks to use the principles of empiricism, skepticism, relativism, objectivity, ethical neutrality, parsimony, accuracy, and precision to assess a particular theoretical explanation.

In the previously mentioned formula, "empiricism" is defined as seeking answers to questions through direct observation. "Skepticism" is the search for disconfirming evidence and the process of continuing to question the conclusions and the evidence that are found. "Objectivity" mandates that conclusions are based on careful observation that sees the world as it really is, free from personal feelings or prejudices. Criminologic researchers often acknowledge that total objectivity is unattainable but every reasonable effort is made to overcome any subjective interests that might influence research outcomes. This is known as "intersubjectivity." Ethical neutrality builds on objectivity by stressing that the researcher's beliefs or preferences are not allowed to influence the research process or its outcomes. "Parsimony" is the attempt to reduce the sum of possible explanations for an event or phenomenon to the smallest possible number. "Accuracy" requires that observations be recorded in a correct manner exactly as they occurred. "Precision" is specifying the number of subcategories of a concept that are available (Adler & Clark, 2007; Maxfield & Babbie, 2009; Vito, Kunselman, & Tewksbury, 2008).

The Relationship Between Theory and Research

As was discussed in a prior section, the practice of criminal justice is based on theories about the causes of crime and how to respond to them. Criminology is an academic discipline that studies the nature of crime, its causes, its consequences, and society's response to the crime. Criminal justice as an academic discipline tends to focus more on the creation, application, and enforcement of criminal laws to maintain social order. There is so much of an overlap between the two disciplines that within this text we deal with the two as one discipline (as indeed many criminologists and justicians consider them to be). Regardless of the reader's orientation, theory is integral in the development of research. Likewise, theory that has been validated by research is the basis for practice in the criminal justice system.

Theory

Theory suggests how something should be (Bachman & Schutt, 2008; Bickman & Rog, 2009; Bryman, 2008; Kraska & Neuman, 2008; Lavrakas, 2008; Maxfield & Babbie, 2009). Personal ideologies are of no value in criminologic theory unless they can be evaluated scientifically. We define theory as "an attempt to explain why a particular social activity or event occurs." A theory is a generalization about the phenomenon being studied. From this broad theory, more precise statements (concepts) are developed. Specific measurable statements are "hypotheses." It is through observation and measurement that the validity (correctness or ability accurately to predict what it seeks to examine) of a hypothesis is examined. If the hypothesis cannot be rejected, then support for the theory is shown. The method by which the hypothesis is observed and measured is known as "research." The relationship between theory and research may be either inductive or deductive in nature.

Inductive Logic

In the stories by Sir Arthur Conan Doyle his detective hero, Sherlock Holmes, continuously assails Dr. Watson, a man of science, about the merits of "deductive logic." It is through deductive logic that Holmes is said to solve his cases. In actuality, the process that Holmes describes is "inductive logic." In this process the researcher observes an event, makes empirical generalizations about the activity, and constructs a theory based on these activities. Only rarely does Holmes engage in the deduction of which he so highly speaks.

Another example of inductive logic is Sir Isaac Newton's alleged formulation of the Theory of Gravity after observing an apple fall from a tree.

Deductive Logic

Deductive logic begins with a theoretical orientation. The researcher then develops research hypotheses that are tested by observations. These observations lead to empirical generalizations that either support or challenge the theory in question. Had our hero Holmes followed up his theory construction with such observation, then he would have engaged in deduction. The scientific method is based on deductive theory construction and testing. In criminologic research, the distinctions between inductive and deductive logic are often obscured because the two processes are actually complementary. Although described in a circular model (Wallace, 1971), the elements of both inductive and deductive logic may also be viewed as part of a never-ending continuum that begins with theory, which encourages creation of hypotheses, and which in turn calls for observations. The result of observations is generalizations, and the conclusions of the generalizations assist in modification of the theory.

The Purpose of Research

The average college student truly believes he or she knows what it means to conduct research. Many have written a "research paper" either in high school or for a college course. Realistically, few have ever had the opportunity truly to write a research paper because even fewer have ever conducted scientific research.

What Is Research

Research is the conscientious study of an issue, problem, or subject. It is a useful form of inquiry designed to assist in discovering answers. It can also lead to the creation of new questions. For example, a judge wants to know how much effect her sentencing has had on individuals convicted of drug possession, particularly as it compares to another judge's sentencing patterns. She asks that research be conducted that focuses on recidivism of these individuals. The results indicate that 30% of drug offenders sentenced in her court are rearrested, compared to only 20% from the other judge's court. Between the two courts, the judge has discovered that her sentencing does not seem to be as effective. This answered the primary question of the research but it has also created new questions, such as why her methods are not working as well as the other judge's.

Research creates questions, but ultimately, regardless of the subject or topic under study, it is the goal of research to provide answers. One of the more common uses of the term "research" is a description of what a student might be asked to accomplish for a college class. Often, one hears instructors and students refer to the choosing of a topic, using several sources, and writing a descriptive paper on the topic as research. If done thoroughly and objectively, this may actually constitute qualitative research (discussed in detail in Chapter 6). Unfortunately, these "research papers" are too often essays based more on the individual's ideologies rather than on scientific discovery. For the purpose of this text, the emphasis is on empirical research that yields scholarly results.

There are many formal definitions for the term "research." The following is adopted here: research is the scientific investigation into or of a specifically identified phenomenon and is applicable to recognizable and undiscovered phenomena. Therefore, in terms of criminal justice and criminology, related research can be viewed as the investigation into or of any phenomenon linked to any or all aspects of the criminal justice system.

Using this definition, criminal justice and criminologic research is not limited to any one area. Box 1-1 offers just a few of the related topics one might research.

Box 1-1

Applied Research Topics: Some Examples

Policing
    Stress
    Patrol effectiveness
    Use of force
    Job satisfaction
    Community policing
    Citizen satisfaction
Courts
    Types of sentencing
    Plea bargaining
    Race and sentencing
    Jury versus judge verdicts
    Paid versus public defender
Corrections
    Rehabilitation versus punishment
    Community corrections
    Boot camps
    Restorative justice
    Death penalty
Other
    Criminal behavior
    Victims
    Drugs
    Gangs
    Juvenile criminality

Along with the plethora of research topics, there are several methods for conducting research. These include surveys, observation, case studies, and reviewing official records. These methods are discussed in further detail, but first it is important to understand all the underlying characteristics of research. To begin with, criminal justice and criminologic research is often divided into two forms: applied and basic.

Applied Research

Perhaps the most immediately useful type of research in criminal justice is applied research, which is primarily an inquiry of a scientific nature designed and conducted with practical application as its goal. It is the collection of data and its analyses with respect to a specific issue or problem so that the applications of the results can influence change. In essence, applied research provides answers that can be used to improve, change, or help decide to eliminate the focus of study. For example, Hernandez (2009) compared several types of self-administered questionnaires among Hispanic juveniles to identify which one was the best predictor of substance abuse.

Basic Research

Basic research, sometimes referred to as "pure research," is the conducting of scientific inquiries that may not offer or provide any direct application or relevance (Drake & Jonson-Reid, 2008; Dunn, 2009). Instead, it is concerned with the acquisition of new information for the purpose of helping develop the scholarly discipline or field of study in which the research is being conducted. This type of research is more often consistent with criminologic inquiries. The findings from basic research often have little or no applicable usage in the field of criminal justice. However, such research may become the foundation on which is based subsequent applied research and criminal justice policy. It is such research that leads to the development of the criminologic theories that guide the actions of lawmakers, police, courts, and corrections. For example, Dantzker and McCoy (2006) explored what process was being used by the largest 17 Texas municipal police agencies for psychologic preemployment screening of police officer candidates. This information was later used to help create a survey tool for a larger research study but initially was simply information gathering or pure research.

Multipurpose Research

Basic and applied research are vital in the study of crime and justice. Yet, a good portion of the research conducted by criminal justice and criminologic academicians tends to come under a third area of research best labeled as "multipurpose research." Multipurpose research is the scientific inquiry into an issue or problem that could be both descriptive and evaluative. That is, it is between the basic and applied realms. This type of research generally begins as exploratory but is of such a nature that its results could ultimately be applicable. For example, a police chief is interested in the level of job satisfaction among his sworn employees. A job satisfaction survey is conducted that offers a variety of findings related to officers' satisfaction. From a basic perspective, the data may simply describe how officers perceive satisfaction with differing aspects of their jobs, becoming descriptive in nature. However, these same findings could be used to evaluate the police agency by examining those areas where satisfaction is the lowest and leading to efforts to determine how to improve these areas. This is the applied nature of the research. The result is research that is multipurpose.

Whether applied, basic, or multipurpose, research can provide interesting findings about a plethora of problems, events, issues, or activities. Regardless of the strategies used, criminologic and criminal justice research is necessary for understanding crime and criminality and for developing suitable responses.

Types of Research

Before conducting research, one must understand something about research; that is, one must first study how research is correctly conducted. At some point in one's college career or during one's employment, a person may be asked to look into something or research a topic. Often, the individual has no clue where to look, how to begin, or what to look for. Then, once the information is obtained, the person may not understand how the information was found and what it actually means.

The primary reason for studying research is to be able to attain a better understanding of why it was done and how it may be used (Bickman & Rog, 2009; Hagan, 2006; Maxfield & Babbie, 2009). Ultimately, if one does not understand what research is and how it works, one cannot understand the products of research. Therefore, the answer to why one studies research is the same reason as why one conducts research: to gain knowledge. This knowledge may occur in one of four formats or types: (1) descriptive, (2) explanatory, (3) predictive, and (4) intervening knowledge (Bachman & Schutt, 2008; Bryman, 2008; Kraska & Neuman, 2008; Lavrakas, 2008).

Descriptive Research

Knowledge that is descriptive allows one to understand the essence of a topic. Research of this nature helps one gain a better grasp about an issue or problem of which one knows little. For example, women have played some role in criminality in this country for years. Yet, very little is known about women and criminality, especially with respect to certain types of crime (e.g., organized crime). To understand better what role women have played, a descriptive study might be conducted. Descriptive knowledge is a very common result of criminal justice and criminologic research. Although the results might be very informative, what can be done with this knowledge is often limited.

Explanatory Research

Explanatory research tries to tell why something occurs, the causes behind the event. This research can be very important when trying to understand why certain types of individuals become serial murderers, or what factors contribute to criminality. Knowing the causes behind something can assist in finding ways to counteract the behavior or the problem. For example, research focusing on gang membership may help to explain why some individuals and not others join gangs. This information could assist in deterring potential future gang members. Ultimately, this type of research may provide answers to questions of how and why.

Predictive Research

Knowledge that is predictive in nature helps to establish future actions. This type of research can be useful to all criminal justice practitioners. For example, if research indicates that a large percentage of juveniles placed in boot-camp environments are less likely to become adult offenders, these results could be used in the future sentencing of juvenile offenders. Conversely, if boot camps are shown to have little or no effect, other alternatives may then be explored. Predictive knowledge gives some foresight into what may happen if something is tried or implemented. Because one of the concerns of criminal justice is to lower criminality, predictive knowledge could be quite useful in attaining this end.

Intervening Research

Intervening knowledge allows one to intercede before a problem or issue gets too difficult to address. This type of research can be quite significant when a problem arises that currently available means are not properly addressing. Research on the effectiveness of certain community policing programs is a good example of intervening research. It can demonstrate whether a specific type of action taken before a given point provides the desired results.

Whether the research is descriptive, explanatory, predictive, or intervening, it is important to understand what research is and how it is valuable. If one fails to study research in and of itself, then all research is of little value. This becomes especially true for the criminal justice and criminologic academic or practitioner who wants to make use of previously conducted research or to conduct his or her own research. It is important to have a grasp of what research is and why it is conducted, before one can actually conduct research.

Why Research Is Necessary

There are a number of specific reasons for conducting criminal justice or criminologic research. Ultimately, the reason is because it is of interest to the researcher. Three primary reasons include (1) curiosity, (2) addressing social problems, and (3) the development and testing of theories.

Curiosity

Being curious is wanting to know about an existing problem, issue, policy, or outcome. For example, in the early 1990s Dantzker and Ali-Jackson were interested in what effect a course might have on students' perceptions of policing. A primary reason for this research was the curiosity of one of the researchers who taught police courses and wanted to see whether there were any differences between perceptions at the beginning and at the end of the course.

Social Problems

The most salient social problem related to criminal justice is crime. Who commits it? Why do they act as they do? How do they do it? These are questions of interest for many criminal justice and criminologic practitioners and academics. Concern over the effects of crime on society only adds further reason to conduct related research. This research can help identify who is more likely to commit certain crimes and why, how to deal better with the offenders and the victims, and what specific parts of the system can do to help limit or even alleviate crime. As a major social problem, crime provides many reasons for research and avenues for exploration.

Theory Testing

Linked more closely with pure criminologic research, theories provide good cause to conduct research. The relationship between theory and research was discussed previously in this chapter. Theory construction is discussed in detail in Chapter 4.

Factors That Influence Research Decisions

Regardless of why the research is conducted, one must be cognizant of factors that can influence why and how it is conducted (Bachman & Schutt, 2008; Lavrakas, 2008). These factors should be identified and carefully considered before starting the research. There are three main influential factors: (1) social and political, (2) practicality, and (3) ethical considerations (Bickman & Rog, 2009; Bryman, 2008; Hagan, 2006; Kraska & Neuman, 2008; Maxfield & Babbie, 2009).

The social and political influences are often specific to the given research. Criminology and criminal justice as social sciences are greatly influenced by social and political events taking place in society. For example, race and ethnicity, economics, and gender might be influential on research about prison environments. Research on whether a particular law is working might have political ramifications. The inability of the criminal justice system to address problems identified by research may not be caused by the lack of system resources but rather by a lack of social desire or political will.

When it comes to conducting research, practicality can play an extremely important role. Economics and logistics are two elements of practicality. How much will the research cost? Can it be conducted in an efficient and effective manner? Would the benefits that are anticipated justify the social, political, and economic costs? Would limited resources be taken from other areas? These are just some of the questions of practicality that could influence the conducting of research and the subsequent uses of that research.

Because ethics plays an important role in conducting research, a more in-depth discussion is offered in Chapter 2. It is briefly noted here that there are three ethical considerations of importance: (1) invasion of privacy, (2) deception, and (3) potential harm. Within a free society citizens jealously protect their rights to privacy. These rights are not just expected by citizens but are protected by law. Deception can have adverse effects not only on the research findings but also on the individuals who were deceived by the researcher. Harm to others, especially to those who did not willingly accept such risks, must be avoided. Each of these considerations is explored later in greater detail.

Whatever the reason, researchers must be aware of the influences that have led to the research and those that might affect the research outcomes. Each could be detrimental to the outcome of the research.

How Research Is Done

Whether the research is applied or basic, qualitative or quantitative (to be discussed in later chapters), certain basic steps are applicable. There are five primary steps in conducting research: (1) identifying the research problem, (2) research design, (3) data collection, (4) data analyses, and (5) reporting of results. Each of these is given greater attention later in the text, but a brief introduction here is appropriate.

Identifying the Problem

Before starting a research project, one of the most important steps is recognizing and defining what is going to be studied. Identifying or determining the problem, issue, or policy to be studied sets the groundwork for the rest of the research. For example, embarking on the study of crime can be too great an undertaking without focusing on a specific aspect of crime, such as types, causes, or punishments. Therefore, it is important first to specify the target of the research. Doing this makes it easier to complete the remaining stages.

Research Design

The research design is the blueprint, which outlines how the research is to be conducted. Although the design depends on the nature of the research, there are several common designs used in criminal justice and criminology. Various designs are presented in this section, and discussed in detail in later chapters.

Survey Research

One of the most often used methods of research is surveys. This approach obtains data directly from the targeted sources and is often conducted through self-administered or interview questionnaires. This type of research generally allows for use of a large sample.

Field Research

Field research is when researchers gather data through firsthand observations of their targets. For example, if a researcher wanted to learn more about gang membership and activities he or she might try "running" with a gang as a participant-observer. This is one of the more time-consuming and limiting designs.

Experimental Research

Experimental research is also observational research. Unlike field research, however, observational studies involve the administration of research stimuli to participants in a controlled environment. Because of ethical and economic concerns, this kind of experimental research is conducted less frequently in criminal justice than are other research strategies.

Life Histories or Case Studies

One of the simplest methods of research in criminology and criminal justice is the use of life histories or case studies. Often these studies require the review and analysis of documents. This type of research might focus on violent behavior where the researcher investigates the lives of serial murderers to try and comprehend why the persons acted in a particular manner.

Record Studies

Another research design is where the researcher evaluates and analyzes official records for relevant data. Previously collected data can be a time saver, but the timeliness of the data may come into question. For example, to determine patterns and influences of robbery, the research design might use data from Uniform Crime Reports.

Content Analysis

In this research design documents, publications, or presentations are reviewed and analyzed. A researcher might review old documents to determine how crime events were publicized in a prior century or may monitor current television broadcasts to assess how the entertainment media influences public perceptions of crime. As another example, to identify the qualifications sought for police chiefs, a researcher could review published advertisements for the position of police chief.

These designs offer a variety of options. There are other possible design methods, which are discussed later in the text. Ultimately, the design used depends on the nature of the study.

Data Collection

Regardless of the research design, data collection is a key component. A variety of methods (discussed in more detail later in the text) exist. They include surveys, interviews, observations, and previously existing data.

Data Analysis

How to analyze and interpret the data is more appropriately discussed in another course, perhaps one focusing on statistics. However, it is an important part of the design and cannot be ignored. The most common means for data analysis today is through the use of a computer and specifically oriented software.

Reporting

The last phase of any research project is the reporting of the findings. This can be done through various means: reports, journals, books, or computer presentation. How the findings are reported depends on the target audience. Regardless of the audience or the medium used, the findings must be coherent and understandable or they are of no use to anyone.

There is one last area worthy of a brief discussion. Information has been offered on why and how to conduct research, but when is it inappropriate to conduct research? Often it seems that research is conducted with little concern as to the appropriateness of the research. Failing to consider this might render the findings useless. Therefore, it has been suggested that the prospective researcher be able to answer the following questions with a negative response (Eck & La Vigne, 1994):

1. Does the research problem involve question(s) of value rather than fact?
2. Is the solution to the research question already predetermined, effectively annulling the findings?
3. Is it impossible to conduct the research effectively and efficiently?
4. Are the research issues vague and ill defined?

If the answer to any of these questions is yes, the research in question should be avoided.

Summary

Conducting criminologic research goes beyond looking up material on a subject and writing a descriptive paper. Before conducting research, one must understand what it is, why it is, and how it might be conducted. For the purposes of this text criminal justice and criminologic research are defined as the investigation into or of any phenomenon linked to any or all aspects of the criminal justice system. The type of research conducted can be applied, basic, or multipurpose. A primary reason for conducting research is to gain knowledge, which can be descriptive, explanatory, predictive, or intervening in nature. Studying research is required to understand better the results offered.

All research tends to follow five basic steps: (1) recognizing and defining a problem, issue, or policy for study; (2) designing the research; (3) collecting data through survey, interviews, observation, or examining previously collected data; (4) analyzing the data; and (5) reporting the findings. Finally, it is important to determine whether it is prudent to conduct the research in question.

Research plays a very important role in criminal justice and criminology. It brings questions and answers, debates, and issues. Knowing what it is, why it is done, and how it can be accomplished is necessary if one is to study crime and criminal behavior.

METHODOLOGICAL QUERIES

For the duration of the text, the methodological queries will be linked to the following:

You are one of several individuals being considered to conduct a study to measure job satisfaction among correctional officers in your county's jail facility. Prior to awarding the research contract, each candidate has to demonstrate he or she is the right person for the job. The process requires you to correctly answer all the questions at the end of each chapter.

1 The sheriff tells you that he believes his employees are generally satisfied with the department simply based on his casual observations. You know that casual observations can be flawed for a number of reasons. How would you explain this to him?

2 Having explained to the sheriff how casual observation can be error prone, you suggest that using the scientific method can account for errors in casual observation. Explain how.

3 The sheriff insists that it would be easy to deduce the level of job satisfaction of his employees. What are the differences between inductive and deductive logic?

4 If given the opportunity to conduct the proposed research, what type of research would you be conducting: basic, applied, or multipurpose research?

5 What type of research design might you use? Why?

6 Describe the steps you would take to complete the research.

Research and Ethics

2

What You Should Know!

Conducting research can be simplistic and uncomplicated. The previous chapter set the foundation for understanding what it means to conduct research in criminology and criminal justice. However, before describing how to do the research, it is important that the prospective researcher be aware of the ethical aspects and apply appropriate ethics. From this chapter the reader should be able to do the following:

1. Define what is meant by ethics and explain its importance to criminologic research.
2. Present and discuss the various characteristics of ethical problems in criminologic research.
3. Explain how the researcher's role influences and is influenced by ethical concerns.
4. Discuss the various ethical considerations presented.
5. Describe the relationship that exists between ethics and professionalism including a "code of ethics."
6. List and describe the four ethical criteria.
7. Present and discuss the five reasons why confidentiality and privacy are important research concerns.
8. Describe the impacts of institutional review boards and research guidelines (such as those mandated by the National Institute of Justice) on criminologic research.

Ethics

Ethics as discussed in this chapter refers to doing what is morally and legally right in the conducting of research. This requires the researcher to be knowledgeable about what is being done; to use reasoning when making decisions; to be both intellectual and truthful in approach and reporting; and to consider the consequences, in particular, to be sure that the outcome of the research outweighs any negatives that might occur. Using this approach, ethical decisions are much easier.

Criminology and criminal justice are virtual playgrounds of ethical confrontations. There is no aspect of them in which ethical questions or dilemmas do not exist, including research. This is particularly true when the research is of an applied nature. The ethical issues encountered in applied social research are subtle and complex, raising difficult moral dilemmas that, at least on a superficial level, seem impossible to resolve. These dilemmas often require the researcher to strike a delicate balance between the scientific requirements of methodology and the human rights and values potentially threatened by the research (Bryman, 2008; Creswell, 2008).

Criminal justice and criminologic research almost always involve dealings with humans and human behavior. It is prudent to be aware of the characteristics associated with ethical problems in social research. Although there does not seem to be a consensus as to what these characteristics are, and there is no comprehensive list, the following have been identified as recognizable characteristics of ethical problems (Bachman & Schutt, 2008; Dunn, 2009; Kraska & Neuman, 2008; McBurney & White, 2007):

1. A single research problem can generate numerous questions regarding appropriate behavior on the part of the researcher.
2. Ethical sensitivity is a necessity but is not necessarily sufficient to solve problems that might arise.
3. Ethical dilemmas result from conflicting values as to what should receive priority on the part of the researcher.
4. Ethical concerns can relate to both the research topic and how the research is conducted.
5. Ethical concerns involve both personal and professional elements in the research.

When dealing with humans, ethics plays an important role. It all begins with the researcher's role.

The Researcher's Role

Contrary to popular belief, the justician or criminologist who conducts research is considered a scientist. Ignoring the distinctions made between a natural scientist and a social scientist, both are scientists who are governed by the laws of inquiry (Kaplan, 1963). Both require an ethically neutral, objective approach to research. As mentioned in Chapter 1, ethical neutrality requires that the researcher's moral or ethical beliefs not be allowed to influence the gathering of data or the conclusions that are made from analyzing the data. Objectivity means striving to prevent personal ideology or prejudices from influencing the process. As can be seen, the two have a similar concern: maintaining the integrity of the research. In addition to these concerns, the researcher, whether a nuclear physicist or a criminologist, must also ensure that the research concerns do not negatively impact on the safety of others.

The researcher's role often coexists and at times even conflicts with other important roles, such as practitioner, teacher, academic, scholar, and citizen. This meshing of roles can often cause the researcher to lose objectivity in his or her approach to the collection, analysis, and reporting of the data. In particular, there are the concerns over the individual's morals, values, attitude, and beliefs interfering with completing an objective study.

Individuals are raised with certain ideals, identified as morals and values. What those are is commonly reflected in one's attitudes and behaviors. Weak or strong morals and values can affect how one conducts research. For example, individuals raised to believe that success is very important, regardless of the costs, might regard the "borrowing" of someone else's research efforts and passing them off as their own as acceptable; or they might accept the manipulation of data to gain more desirable results. An even more repugnant scenario is one in which the researcher continues with his or her research despite knowing that to do so will cause physical harm or emotional anguish for others. In each of these cases, ethically the decisions are wrong.

Because the researcher's role is intertwined with other roles, ethics becomes even more difficult to manage. Ultimately, it is up to the individual to decide the importance of personal ethics. However, this is just one aspect of ethics in research.

Ethical Considerations

Conducting research in and of itself can be problematic. Accessibility, funding, timing, and other factors may all impose problems. The reality is there can be ethical concerns at every step of the research process (Bickman & Rog, 2009). With this in mind, the considerations discussed next should not be viewed as more important at any one particular time in the process, but rather they apply throughout the research.

Ethical Ramifications

One of the first things to consider is whether the topic to be studied has innate ethical ramifications. Some topics are controversial by their very nature. For instance, the individual interested in gangs might decide that the best way to gain data is to become a participant observer. As such, chances are that the researcher may have to witness or even be asked to participate in illegal activity. Ethically as well as legally, this information should be given to the police, but doing so might jeopardize the research. Although it is apparent what decision should be made (the research should be adjusted to avoid such a dilemma or possibly even abandoned outright), the right one is not always made simply because of how important the research is perceived to be to the individual. Therefore, before embarking on a research topic, the ethical implications of the research itself must be addressed.

Harm to Others

Another consideration is what effects the research might have on the research targets. When the research involves direct human contact, ethics plays an important role. Whether the targets are victims, accused offenders, convicted offenders, practitioners, or the general public, a major consideration is whether the research might cause them any harm. Harm can be physical, psychologic, or social.

Physical harm most often can occur during experimental or applied types of research, such as testing new drugs or weapons. Psychologic harm might result through the type of information being gathered. For example, in a study of victims of sexual assault, the research might delve into the events before, during, and after the assault. This line of questioning may inflict more psychologic harm in addition to that which already exists as a result of the assault. Finally, social harm may be inflicted if certain information is released that should not have been. Consider a survey of sexual orientation among correctional officers where it becomes public knowledge as to who is gay. This information may cause those individuals to be treated differently, perhaps discriminated against, causing sociologic harm. It is important that the researcher consider what type of harm may befall respondents or participants before starting the research.

Privacy Concerns

The right to privacy is another ethical consideration. Individuals in America have a basic right to privacy. In many cases, research efforts may violate that right. How far should individuals be allowed to pry into the private or public lives of others in the name of research? Ethically speaking, if a person does not want his or her life examined, then that right should be granted. All persons have a right to anonymity. However, there are a variety of documents accessible to the public in which information can be gathered that individuals would prefer to be unavailable, such as arrest records, court dockets, and tax and property records. The ethical question that arises here is whether a person should have the right to consent to access to certain types of information in the name of research. Giving consent in general is a major ethical consideration.

Particularly in survey research it is common for the researcher either to ask for specific consent from the respondents or at least acknowledge that by completing the survey, the respondent has conferred consent. Normally, this only requires having the individual sign an informed consent form or for the instructions to indicate that the survey is completely anonymous, voluntary, and that the information is only being used for the purpose of research (Figure 2-1).

Voluntary Participation

As should have been noted in the previous example, not only did the researchers seek to obtain consent, they also informed prospective respondents that participation was voluntary. Too frequently criminologic researchers require their subjects to sign consent forms but (particularly within institutional settings, such as military organizations, schools, and prisons) neglect to inform them that their participation is voluntary. In fact, in these environments, participation is often coerced. Not all research must use voluntary participation, but it is stressed that there must be valid reasons that can be given showing that the knowledge could not otherwise be reasonably obtained and that no harm will come to the participants from their compulsory involvement.

Regardless of the fact that the research was not intrusive and could cause no harm to the respondents, informed consent was required. The rule of thumb in these situations is if there is any doubt as to whether the research could be in any way construed to be intrusive, then consent should be obtained from the subjects. It is also best to assure them that their participation is voluntary and they may choose not to take part in the study.

Within the academic setting, informed consent and voluntary participation do not seem to be an unusual requirement. To ensure that informed consent is provided, and to judge the value and ethical nature of the research, many universities have an Institutional Review Board (IRB). The IRBs exist as a result of the Code of Federal Regulations, Title 28 Judicial Administration, part 46, which specifies all aspects of the IRB including membership, functions and operations, reviewing the research, and criteria for IRB approval (http://www.hhs.gov/ohrp/assurances/index.html).

Established primarily for the review of research, usually experimental or applied, dealing directly with human subjects, university IRBs often extend their review over any type of research involving human respondents (survey or

Pre-employment psychological screening tools for police candidates: Psychologists' choices and reasons

Before you continue with this short online survey, please read carefully the following consent form and click the **"I CONSENT"** button at the end to indicate that you agree to participate in this data collection effort. It is very important that you understand that your participation in this survey is voluntary and that the information you share is private.

You were selected to participate in this through a random selection of members of the APA who designated clinical psychology as their specialty and/or because of your membership in the Psychology section of the IACP. The survey includes a series of closed ended questions asking you about choices and reasons for identifying specific psychological tools for use as part of pre-employment screening of police officer candidates. **The findings from this survey will be used to produce articles informing interested parties of the results and recommendations formed. The overall intent will be to start a serious dialogue toward recognizing police psychologists as specialists for conducting psychological screening of police officer candidates, as well as the development of a standardized set of testing protocols.**

There are no right or wrong responses to this survey and the survey will take approximately 15 minutes for you to complete. Your consent to participate in this survey requires that you carefully read and agree to the following:

Privacy: The information that you provide via this survey will be kept private except as otherwise required by law. Any identifying information will not be disclosed to anyone but the researchers conducting this evaluation and will be kept in locked files separate from the data collected. However, the potential identifying data being collected is of the nature where it will be nearly impossible to identify any particular individual. The information reported will not contain any identifying information.

Risks: Completing this survey poses few, if any, risks to you. You may choose to cease input of information at any time or to not answer a question, for whatever reason.

Your participation is voluntary. Refusal to participate involves no penalty or adverse consequences. If you consent to participate in this survey here are some additional things you should know:

- You may stop your input of data at any time without penalty or consequence.
- You may choose to not answer a question at any time without penalty or consequence.
- You may contact the researcher or his faculty supervisor, Dr. Sandra Mahoney, sandra.mahoney@waldenu.edu, with any questions that you have about the research, during or after you have completed the survey.
- There is no compensation being offered to participate.
- I encourage you to print a copy of this consent for your records.

Contact information: If you have any concerns about your participation in this survey or have any questions about the evaluation, please contact M. L. Dantzker, Principal Investigator at mldcjc@att.net or at 956 682-9364. Walden University's approval number for this study is **05-18-09-0327204** and it expires on **May 17, 2010.** Please click the "I CONSENT" box below to participate in the study.

Thank you.

FIGURE 2-1 Informed consent example.
Source: Reproduced from Dantzker, M. L. *Psychologists' Role and Police Pre-Employment Psychological Screening.* ProQuest Company, 2010.

otherwise). Although having to attain IRB approval can be somewhat frustrating, it is a useful process because it helps to reaffirm the researcher's perceptions and beliefs about the research and can help identify prospective ethical problems. Also, reviewers may see problems overlooked by the researcher. It is better to err on the side of caution.

The process generally is not that difficult. It usually requires the researcher to submit basic information about the proposed research, often in a format designed by the university. Appendix A contains an example of a request submitted to an IRB for approval. Although not all IRBs make use of the same format, the information required is similar across institutions.

Informed consent is valuable because it is important that research targets are allowed the right to refuse to be part of the research. Although in survey research consent may not be a major problem (because permission can be written into the documents), it does raise an interesting dilemma for observational research (when the researcher may not want the subjects to know they are being observed). The ethical consideration here is that as long as the subjects are doing what they normally would be doing and the observations do not in any way directly influence their behavior or harm them, it is ethically acceptable.

Deception

Some types of research (particularly field research that requires the researcher to in essence "go undercover") cannot be conducted if the subjects are aware that they are being studied. Such research is controversial and must be carefully thought out before it is undertaken (Vito, Kunselman, & Tewksbury, 2008). All too often the deception is based more on the researcher's laziness or bias rather than a real need to deceive. For example, a researcher is interested in studying juvenile behavior within the confines of a juvenile facility. Rather than explain to administrators and the subjects what he or she is doing, the research is conducted under the guise of an internship or volunteer work.

Depending on the type of research, there are always some ethical considerations. What is interesting is that the science of research itself is viewed as ethically neutral or amoral. The ethical dilemmas rise from the fact that researchers themselves are not neutral. This fosters the need for regulation in the conducting of research so that it does meet ethical standards (Fowler, 2009; Gavin, 2008).

The Professionalism of Research

According to Merriam-Webster's Online Search (2009), a professional is one whose "conduct, aims, or qualities . . . characterize or mark a profession or a professional person" (http://www.merriam-webster.com/dictionary/professionalism).

A profession is defined as "**a:** a calling requiring specialized knowledge and often long and intensive academic preparation **b:** a principal calling, vocation, or employment **c:** the whole body of persons engaged in a calling" (http://www.merriam-webster.com/dictionary/profession). Research in itself is a profession, and when mixed with other professions there is an even greater need to conduct business in a professional manner. This often means that the profession has established a code of ethics.

Many professions have support of written codes of ethics for research (i.e., The American Psychological Association). However, although criminal justice and criminology do not have a globally applicable code, the Academy of Criminal Justice Sciences, an organization to which many academic researchers are members, did develop a code of ethics for its members that includes a section on researcher ethics (Appendix A). Furthermore, although there seems to be no universal code of ethics with respect to research, grant-funded research is more likely to have ethical constraints imposed. For example, a popular source of funding for criminal justice and criminologic research is the National Institute of Justice (NIJ). NIJ has developed its own "code of ethics" to which all grant recipients must agree. The NIJ is very specific in its guidelines, especially with respect to data confidentiality and the protection of human subjects (Figure 2-2).

Ethical Research Criteria

Even though there is no universally recognized research code of ethics, there are some specifically identified criteria that, when applied or followed, assist in producing ethical research. These criteria, discussed next, include avoiding harmful research, being objective, using integrity, and protecting confidentiality.

Avoiding Harmful Research

The goal of research is to discover knowledge not previously known or to verify existing data. In many instances this can be done without ever having to inflict any undue stress, strain, or pain on respondents (i.e., historical or survey research). Unfortunately, at times research can be physically or emotionally harmful. The ethical approach is to avoid any such research regardless of how important its findings might be unless it can be shown that good from the information far outweighs the harm (an eventuality that is rare even in criminologic research).

Being Objective

Biases can be detrimental to a research project. One such bias deals with objectivity. Assume you do not like drinkers, that you perceive them as weak willed and careless. Your research deals with individuals convicted of driving while

All NIJ employees, contractors, and award recipients must be cognizant of the importance of protecting the rights and welfare of human subject research participants. All research conducted at NIJ or supported with NIJ funds must comply with all Federal, U.S. Department of Justice (DOJ), Office of Justice Programs, and NIJ regulations and policies concerning the protection of human subjects and the DOJ confidentiality requirements.

Why is it important for those of us involved in research to care about these requirements? Fulfilling our obligations under these regulations is important for several reasons other than just being in compliance with the regulations and processing the research award, including:

- Following these procedures provides research subjects protection from harm that might result from their participation in research.
- Complying with these procedures (e.g., IRB review, informed consent, confidentiality concerns) improves the overall quality of the research we conduct and the data used in the analysis.
- Consideration of the confidentiality and human subject issues and compliance with the rules will allow us to continue to conduct difficult research on important societal problems and to provide a scientifically informed basis for making important public policy decisions.
- The codes of conduct and ethical standards of our profession to which we adhere require the dutiful protection of human research subjects and confidentiality.
- Many of these concepts have longstanding associations with other fundamental aspects of our society (e.g., belief in individual rights, representative government), and fulfilling our obligations defines us as a society and a nation.

NIJ policy provides for the protection of the privacy and wellbeing of individuals who participate in NIJ research studies under two different, but philosophically related, sets of regulations:

Human Subjects Protection ("The Common Rule")
Confidentiality and Privacy

Figure 2-2 NIJ human subjects and privacy protection.
Source: Reproduced from "Human Subjects and Privacy Protection," National Institute of Justice, 2010.

intoxicated. You are interested in their reasons for driving while impaired. The chances are good that if you allow your personal feelings against drinkers to guide you in your research, the results will be skewed, biased, and subjective. It is important, for good ethical research, to maintain objectivity. Of course, being objective is just one important characteristic of the ethical researcher.

Using Integrity

The last thing a researcher wants is for the results not to meet expectations. Sometimes, because of how important the research is perceived to be, there may be a tendency to manipulate the data and report it in a manner that shows the research was successful; that is, put a positive spin on an otherwise negative result. This is especially possible when the research is evaluative and its results could influence additional funding for the program being evaluated. When faced with this dilemma, because of the desire not to jeopardize the program's future or to improve future chances for research, the researcher may not report the true findings. This is extremely unethical, but unfortunately, may be more commonplace than one would like to believe. The ethical researcher accepts the findings and reports them as discovered.

Protecting Confidentiality

One of the biggest concerns in conducting research is the issue of confidentiality or privacy. As it has been suggested, privacy and confidentiality are two ethical issues that are crucial to social researchers who, by the very nature of their research, frequently request individuals to share with them their thoughts, attitudes, and experiences.

Because a good portion of criminal justice and criminologic research involves humans, chances are great that sensitive information may be obtained in which other nonresearch efforts might be interested. For example, conducting gang research where street names and legal names are collected perhaps along with identifying tattoos, scars, and so forth, and voluntary statements of criminal history. This information is extremely valuable to a police agency. Ethically, that information must remain confidential.

Reasons for Confidentiality and Privacy

Overall, five reasons have been identified as to why confidentiality and privacy are important in research (Adler & Clark, 2007; Kline, 2009; Maxfield & Babbie, 2009):

1. Disclosure of particularly embarrassing or sensitive information may present the respondent with a risk of psychologic, social, or economic harm.
2. Sensitive information, if obtained solely for research purposes, is legally protected in situations where respondents' privacy rights are protected.
3. Long-term research may require data storage of information that can identify the participants.
4. The courts can subpoena data.
5. Respondents may be suspicious as to how the information is truly going to be used.

The bottom line is that confidentiality and privacy must be maintained. There are two methods of accomplishing this: physical protection and legal protection. Physical protection relates to setting up the data so that links cannot be made between identifying information and the respondents. Reducing who has access can also aid in protecting the data. Legal protection attempts to avoid official misuse. Researchers are aided with this by an amendment to the 1973 Omnibus Crime Control and Safe Streets Act, better known as the "Shield Law," which protects research findings from any administrative or judicial processes. As noted previously, funded research through such organizations as NIJ or the National Institutes of Health is overseen by organizational regulations. Unfortunately, these guidelines do not completely protect the data, leaving researchers responsible for gathering the data in a manner that best protects the respondents.

By simply meeting the four suggested criteria, a researcher can avoid many ethical problems. However, perhaps the best way to avoid ethical problems is to conduct research using a method that does not compromise ethical standards: research that is legal, relevant, and necessary.

Summary

The simple act of research, especially when it involves humans, creates a plethora of possible ethical dilemmas. Because ethics is important to professions, researchers need to be cognizant of several ethical considerations. These include determining whether the topic itself is ethical, what harm or risk is involved to respondents, and confidentiality and privacy. There are federal guidelines for protecting individuals' privacy and for obtaining their consent, which in the university setting is often reinforced through an IRB. The key to ethical research is a professional approach. Some professions have created a code of ethics applicable to research. Although criminal justice and criminology do not have one specific to the discipline, a major criminal justice organization has established such a code for its members. However, there are four criteria that when followed, alleviate the need for such a code: (1) avoid conducting harmful research; (2) be objective; (3) use integrity in conducting and reporting the research; and (4) protect confidentiality.

METHODOLOGICAL QUERIES

1 Due to the nature of politics in your county, the sheriff is concerned that the ethics behind the research may be called into question. How would you explain to him what is meant by ethics? How do you assure him of its importance to criminological research?

2 Although the sheriff understands the ethics with regard to politics, he is not clear on the ethical problems that could arise conducting the proposed research. You must present and discuss the various characteristics of ethical problems in criminological research. What do you tell him?

3 Because you live in the county, know some of the correctional officers, and even voted for the sheriff, there are some who might question whether your role as researcher

has ethical implications. To ensure this will not be a problem, you must demonstrate how the researcher's role influences and is influenced by ethical concerns. What do you say or do?

4   One way you may respond to the previous question is to list and describe the four ethical criteria. Explain how you would link the criteria to the proposed research.

5   A major concern for the sheriff is the need to ensure and maintain confidentiality and privacy. Present and discuss the five reasons why this is necessary and how it could be accomplished.

The Beginning Basics

3

What You Should Know!

For established researchers, what to study and how to study is simple. However, for most student researchers a consistent issue is having difficulty in choosing what to research. A major question one might ask is, where do I start? Furthermore, choosing a topic is merely the first step. Other issues include developing the research question, a statement explaining what it is the research is to accomplish. Forming the research question leads to the formation of hypotheses and the identification of variables. Therefore, from this chapter the reader should be able to do the following:

1. Discuss the issues that should be considered in selecting a research topic.
2. Present and describe the three purposes of research.
3. Describe what a literature review is and the sources that are available for such a review.
4. Compare and contrast the various writing styles used by criminologists.
5. Define what an article critique is and discuss its details.
6. Define what is meant by the research question. Give an example of a research question.
7. Define hypothesis and describe the types of hypotheses.
8. Define variable and describe the types of variables.

Getting Started

In the previous two chapters we discussed why research is necessary and the importance of research ethics. These are important topics that warrant serious consideration. However, if the reader is charged with writing an empirical research paper one may find that there is still a great deal needed to know to complete such a project satisfactorily. This chapter presents and discusses a number of issues that must be considered when starting a research project.

One of the most difficult aspects of any endeavor is to begin, and research is no exception. By its very nature, the process by which a research problem is selected is a creative process. This process may require thinking about various ideas and issues, asking questions, and determining whether the question may actually have answers (Bryman, 2008; Creswell, 2008; Drake & Jonson-Reid, 2008).

Picking a Topic

Before beginning the research project, one must first answer this: what should I study? Within the fields of criminology and criminal justice there are numerous research topics available. All one has to do is pick a topic; however, that is not as easy as it may seem.

The beginning of any research project must focus on what is to be studied. Defining the problem is viewed as the most important stage of the research (Fowler, 2009; Gavin, 2008; Hagan, 2006). To start, it should be something of personal interest. If there is no interest in the topic from the start, one will be tired of it before it is completed. It must also comply with any topic restrictions imposed by the individuals or organization for which the research is being conducted. For example, the authors frequently restrict the topics that their students may research to avoid emotional diatribes on controversial issues better left to experienced researchers. In addition, before the topic is chosen, one should consider (1) what currently exists in the literature, (2) any gaps in theory or the current state of art, (3) the feasibility of conducting the research, (4) whether there are any policy implications, and (5) possible funding availability (Fowler, 2009; Gavin, 2008; Hagan, 2006; Maxfield & Babbie, 2009; McBurney & White, 2007; Vito, Kunselman, & Tewksbury, 2008).

There are a number of studies about community policing. Many of them suggest and support the effectiveness of community policing. What seem to be lacking are studies that explain why it may be successful in some places and a failure in other places. Thus, a gap in theory exists and needs to be filled. Finding such gaps in the literature can assist in choosing a topic.

Once an interesting and intriguing topic is found, the feasibility of conducting the research must be considered. Feasibility is primarily linked to logistics (e.g., Is a sample accessible? Does a data collection instrument exist or must one be designed?). Sometimes the topic may be a very good choice for researching, but is not feasible to attempt. Before finalizing the research topic, the prospective researcher needs to be sure that the study can actually be accomplished.

Because of the popularity of some topics, such as job stress, capital punishment, sentencing disparity, and community policing, there is usually a wealth of information available to help build a research base. However, there may be times when the topic is legitimate, but little information exists in the literature to support the research question. This should not stop one from going forward with the research. If the findings can be validated (shown to meet scientific rigor and supported by ensuing research), it may become a new and significant contribution to the literature, the discipline, and subsequently the practice of criminal justice or criminology.

Choosing a research topic that could have policy implications can be very useful. For example, examining job satisfaction or examining the effect education has on promotion. The results of either study could have an effect on policy and procedure.

Finally, although funding may not be applicable to many students' research efforts, it should be taken into consideration. One of the most popular means of funding research is through internal and external grants. Many universities offer both students and faculty opportunities to apply for internal grants that at least allow the person to start the research project and may help offset personal costs. Ultimately, many researchers seek external grants. However, these are usually not sought by undergraduate students. Regardless of where the funding may come from, it is important to establish whether any monies are needed and from where they may be sought.

The Purpose of the Research

Once consideration is given to the previously mentioned issues, the research can begin, but to address these issues, one must have an idea of what the research will cover. Perhaps the best place to start is to decide what the research should accomplish. As discussed in Chapter 1, there are three possible expectations or accomplishments: (1) exploring, (2) describing, and (3) explaining (Gavin, 2008; Hagan, 2006; Maxfield & Babbie, 2009).

Exploring

Most of us are explorers. Our curiosity about things begins at a very early age and should last until death. What we explore and how we explore something changes with age and time. Much of the exploration occurs either accidentally

FROM THE REAL WORLD

The attitudes of college students can provide interesting insight as to how young adults tend to think of various topics. An area of particular interest among college students is sexuality. A question is whether attitudes differ by major. Because criminal justice students are perceived as a more conservative group, Dantzker and Eisenman (2007) conducted a study among criminal justice students examining their attitudes toward homosexuality, pornography, and other sexual matters. Their findings did tend to support the perception that these students do possess a conservative attitude.

or intentionally. Intentional exploration may well be considered to be a form of research. Thus, when one wants to know more about something, the tendency is to explore the topic.

For example, before buying a new sports utility vehicle, one might read what various magazines say about them, check out prices through various dealers, and talk with current owners of such vehicles. When finished, one should have enough information to make an informed decision: research has been conducted. With respect to criminal justice and criminology, exploration of one's interests is often formal and intentional, and accomplished through some form of research.

Exploratory research provides information not previously known, or about which little is known. It seeks out information about something that is known to exist, but as to why or how, that must be discovered. Therefore, exploratory research offers additional insights about something for which there is awareness but limited knowledge.

Describing

What is the phenomenon? How does it work? What does it do? These are just three of the types of questions answered when conducting research for the purpose of describing. The basic purpose is to be able to describe specific aspects or elements of the topic. Generally, this type of research is informative in nature and is based on something one is already aware of, but knows little about. For example, many people might know what a prison is, yet many really do not know the details about a specific prison. A descriptive study could offer information about the inmates, correctional staff, programs, violent acts, and so forth, which would further enhance what is known about that prison. Descriptive studies are probably among the easier studies to do because the researcher simply needs to explain what he or she sees, hears, or reads with respect to the various elements of the topic. From a purely academic perspective, some literature reviews might serve as a descriptive study.

Explaining

Undoubtedly the most in-depth and difficult purpose for conducting research is to provide an explanation. Explanatory research attempts to analyze and fully understand the concept "why" as it applies to policies, procedures, objects, attitudes, opinions, and so forth. From a criminologic perspective, to understand better why males commit sexual assaults, a study of several convicted rapists where questions are asked that answer "why" might be conducted. For criminal justice, examining acts of police corruption to determine "why" they occur is an explanatory type study.

Sometimes a study may start out exploratory and end explanatory. For example, due to a report of increased criminal behavior, criminologists sought to find out what was going on in a rural high school. They conducted exploratory research to find out whether the criminal behaviors were the result of gang activities. On learning that it was gang related, they could rely on the literature to make recommendations as to how to deal with the problem. Or they might engage in further descriptive research to learn exactly how the gangs were made up and what they were doing. This could then lead to explanatory research as to why gangs were able to get a foothold in the high school, what the impacts of such activities will be, and how to combat these activities.

Although all three reasons for conducting research are valid, it is most often the explanatory reason that many criminologists and justicians pursue. This type of study relies on strong, clear research questions, hypotheses, and variables. Each of these is discussed in greater length later. However, there is still the question of choosing a topic. To demonstrate the myriad of possible topics for study, the following is a short list of potential research examples.

1. Examine the role of criminal justice programs in a given state.
2. Survey education and job satisfaction among police, parole, or correctional officers.
3. Compare and contrast criminal justice educators: practitioner-academic versus pure academic and their work products.
4. Evaluate the relationships between job stress and race and ethnicity.
5. Explore the causes and extent of criminality among high school students.
6. Conduct a cost benefit analysis of private versus state prisons.
7. Research drug usage among adolescents.
8. Evaluate the success of electronic monitoring.
9. Evaluate the effectiveness of community policing.
10. Survey public attitudes toward corporal punishment.

Still, the question lingers on how to go about choosing a topic. This can be aided through personal observations, suggestions from academics and scholars, other students, and the existing literature in criminal justice and criminology. It is through this last means that a large number of research topics are chosen.

Reviewing the Literature

For many researchers, although the choice of an idea or concept to study may, at times, be frustrating, what becomes more frustrating is choosing a topic, and finding either too little or too much information available in the current literature. Neither of these situations should prevent the individual from conducting the research. However, they do make it more difficult to support the need to do research. There is usually a substantial amount of literature to support most topics one might wish to research in criminal justice and criminology. With the extensive number of journals currently available, data from government agencies, and Internet access to information, one can usually collect enough information to support a research effort.

Ultimately, the best way to begin a research effort is to focus on a particular issue, phenomenon, or problem that most interests the individual. In doing so, one must be sure to determine the problem, issue, or phenomenon and organize what is known about it. This is where a literature search and review is valuable, and the best place to begin the search is the library.

Become Familiar with the Library

A first step in conducting a search of the literature is to familiarize oneself with the nearest library. As a criminal justice or criminology student, one should have access to a good library. If not, find out how far it is to a better library and make arrangements to go there. Too frequently students try to use the excuse that "it was not in our library" after a brief perusal of the texts and journals that are available. That is not an acceptable excuse. As an individual who is capable of thinking critically one is expected to solve problems regarding the availability of resources, not to just bemoan them. Familiarize yourself with the layout of your particular library, the search vehicles that are available, and get to know the librarians. There are many more resources available than one may have imagined.

Text and Journal Abstracts

Literature or topical searches can start with the use of a source called an "abstract." Two popular abstracts are the Sociological and Criminal Justice abstracts. With these sources the researcher can look for a particular term (a key word or words, such as "job satisfaction"), concept, or topic to see what has already been published about this subject.

Scholarly Journals

From the abstracts one can go directly to identified journals. Scholarly journals are refereed (meaning that to be published the article appearing in the journal had to pass review by other scholars who were asked by the journal to critique its contents). How critiques are conducted is discussed in a following section. The findings from a perusal of journal articles may not only help determine a topic but also guide research.

Textbooks

For research purposes introductory-level texts are generally not of sufficient depth to use as sources in research. It is better to build from them by going to the sources that they cite within a subject area or to rely on texts devoted to the subject in question. However, they are often excellent starting sources for selecting a research topic. If one is assigned a research project for a certain class, review the contents of the text that is being used in that class to determine if something of personal interest may be revealed.

Social Science Indexes

Annual indexes for journals are another source of information. Today, journals are available on microfiche, as hard copy, or "on-line." Government documents and the Internet are additional sources for helping choose a topic or gathering information to support the research topic.

Internet Searches

A word of caution is appropriate regarding the use of the Internet. Although it is useful for gaining preliminary information about a topic, it does not replace conducting a solid literature review of the subject area. For all its convenience, the Internet has a great deal of information that is unsubstantiated, if not outright erroneous. If the source is not clearly scholarly (e.g., a reputable on-line journal) be cautious about using it as a research reference. The Internet augments the library and social science indexes and abstracts as search vehicles; it does not replace them. Nor does it replace journals and textbooks. Although the Internet may provide legitimate sources, a reference page filled with numerous Web sites instead of text and journal citations is indicative of lazy scholarship on the part of the researcher.

Overall, there are a number of sources from which one can choose a topic or find support for a given topic. Box 3-1 provides a sampling of refereed journals that publish articles in criminology or criminal justice. This is not intended to be viewed as an exhaustive listing. Once a literature search is completed, the research questions can be formulated.

Critiquing the Literature

To conduct a sound literature review (whether as a topical search or in developing the scholarly basis for the research being conducted) one must be able properly to interpret the research being read. This section provides guidance as to what to look for in evaluating other studies. These guidelines are also

Box 3-1

Literature Review: Examples of Sources

Journals
    American Journal of Criminal Justice
    American Journal of Sociology
    British Journal of Criminology
    Canadian Journal of Criminology
    Crime and Delinquency
    Criminal Justice Policy Review
    Criminology Justice Quarterly
    Journal of Contemporary Criminal Justice
    Journal of Criminal Justice
    Police Quarterly
    Prison Journal
    Social Forces
    Social Justice
    Social Problems
    Social Science Quarterly
Compendiums
    Abstracts on Criminology and Penology
    Criminal Justice Abstracts
    Police Science Abstracts
    Psychological Abstracts
    Social Science Index
    Sociological Abstracts
Government Agencies
    Bureau of Justice Statistics
    National Criminal Justice Reference Service (www.ncjrs.org)
    National Institute of Justice (www.ojp.usdoj.gov/nij)

helpful in preparing one's own work for others to review. More detail on preparing work for submission is provided in a later chapter.

Understanding Writing Styles

Scholarly journals and textbooks conform to specific writing styles. The various styles are precisely detailed in publication manuals. Several styles are used in criminal justice and criminology. At some point in the research process one may view Turabian (which uses numbers to indicate citations and footnotes at the bottom of each page); Chicago Style (also uses numbers after citations with endnotes instead of footnotes); American Psychological Association (APA; the most commonly used style in criminal justice and criminology; lists the author and year of publication within the text); and American Sociological Association (similar to APA but may use endnotes for specific details and varies in the format of the references). Occasionally, one may also see Modern Languages

Association style and styles unique to specific law or criminal justice journals (these are often variations of the previously mentioned styles). It is important to know what style is used when critiquing an article or text and even more important to know what style is required when submitting a paper, article, or text for review. For example, the authors require their students to submit all papers in APA format. The best way to become familiar with this format is through the APA stylebook.

Knowing What to Look For

There is a number of things that should be looked for in critiquing another's work. The following have been suggested as main areas one should target for review and understanding (Kline, 2009; Kraska & Neuman, 2008; Lavrakas, 2008).

Abstract

This is nothing more than a brief overview of the article. What is the social issue that was studied? How was it investigated: exploration, description, explanation, or a combination of strategies? Who conducted the research? Who financed it (and is there a conflict of interest)? Was it well written and well organized, and did it have clarity of purpose? Were the findings reasonable based on the research design? What were the conclusions and recommendations?

Clarity of Problem Statement

What was the problem being investigated? Did the literature review support the need for this research? What was the theoretical orientation? What was or were the research hypotheses? Were concepts properly defined? Were the dependent and independent variables identified?

The Literature Review

Did the literature review provide a thorough coverage of the prior research? Were previous studies adequately evaluated and discussed? Was the coverage complete (classic studies relevant to the research problem included and recent research)? Did the literature cited provide a justification for the current investigation?

Methodology Used

What was the research design? Was it clearly developed from the theoretical frame of reference alluded to in the problem statement? Who were the subjects and how were they included in the study? Did the study conform to ethical standards? What was the sampling method and was it adequate for this research? Was the measurement instrument or strategy satisfactory? How were the data analyzed? Were there any adverse effects or limitations because of the research design and the means of analysis being incompatible? Were measures of association and tests of significance clearly indicated and appropriately discussed? Were the measurement techniques valid and reliable?

Findings

Were the findings displayed in a concise and readily understandable manner? How were the data summarized? Were the tables logical and clear? Were the statistical techniques appropriate? Would other statistical techniques have been more appropriate? Did the findings relate to the problem, the method, or the theoretical framework? How did the findings relate to those of the prior research?

Discussion and Conclusions

Are the conclusions reached consistent with the findings that were presented? Are the conclusions compatible with the theoretical orientation presented in the problem statement? Based on the problem statement, the prior research, methodology, and findings, do the conclusions or recommendations make sense? Based on the problem statement and the literature are the conclusions or recommendations of this study a significant contribution to the field of study?

The Research Question

Once a topic has been chosen, the next step is to create the research questions. The research question is a statement answered through the research process. Its focus, which must be clearly stated, should inform readers of the actual purpose of the research. The research question generally is formulated from the research problems or purpose, and may be synonymous with the research problem. Essentially, what the researcher must do is decide what it is he or she is to study and why. It is the "why" that helps form the research question.

The research questions should allow others to gain a clear understanding of why the research was conducted. A well-worded research question should give some indication of the outcomes one might expect at the conclusion of the research. The question should be directly linked to the research problem.

Hypotheses

A hypothesis is a specific statement describing the expected result or relationship between the independent and dependent variables. The development of hypotheses and their linkages with theory and research is discussed further in Chapter 4. The three most common types are (1) the research hypothesis, which is a statement of the expected relationship between variables offered in a positive manner; (2) the null hypothesis, which is a statement that the rela-

FROM THE REAL WORLD

Dantzker (2010) conducting a study of police psychologic preemployment screening offered the research problem as whether there were differences between police psychologists and general clinical psychologists in the evaluative instruments or protocols used in conducting preemployment screening for potential police officers, and do they select those instruments or protocols for reasons of job-specific validity. The work of psychologic screening for potential police candidates is conducted by two types of psychologists: police and general clinical psychologists. Whether there is a difference between the two groups of psychologists in terms of this work has not been examined. One could suggest that if police psychologists do differ in this area from general clinical psychologists, then a question for future research is whether their selections of questionnaires and protocols to conduct such screenings are more applicable than those chosen by general clinical psychologists.

This problem produced the following research question:

Are there differences between police psychologists and general clinical psychologists in the evaluative instruments or protocols used in conducting preemployment screening for potential police officers and do they select those instruments or protocols for reasons of job-specific validity?

After establishing the research questions, the researcher must next explain what specifically is going to be studied and the expected results. This is usually accomplished through statements or propositions referred to as "hypotheses."

tionship or difference being tested does not exist; and (3) the rival hypothesis, which is a statement offering an alternate explanation for the research findings.

The research hypothesis, which is the most common of the hypotheses, is generally a statement that fits the equation, if "X" then "Y." An example is, "If there is an increase in the number of patrol units in a given area, the amount of reported crime increases." Note that the statement suggests a relationship between two variables, patrol units and reported crime, in a manner indicating the belief that more patrol cars will cause there to be more crime reported. Therefore, the research would focus on examining this relationship to disprove the null hypothesis.

The null hypothesis fits the equation, "X" has no relationship with "Y." For the previous research hypothesis, the null hypothesis would read "The increase in patrol units will not increase the amount of reported crime (no relationship exists)." A successful research effort will disprove this, which in essence supports the research effort.

What if the increase in patrol units decreases the amount of reported crime? This would lead to the rival hypothesis that fits an equation, in the previously

FROM THE REAL WORLD

The study by Dantzker from the previous "Real World" offered the following hypotheses:

Hypothesis One: Police psychologists will not differ significantly from general clinical psychologists in their choices of instruments or protocols used for preemployment screening of police officer candidates. Hypothesis Two: Police psychologists will not differ significantly from general clinical psychologists in their reasons for selecting their choices of instruments or protocols used for preemployment screening of police officer candidates.

mentioned circumstance, of the more of "X" the less of "Y." Again, although the general research goal is to support the research hypothesis by disproving the null hypothesis, one should not consider the inability to disprove the null hypothesis as a failure. On the contrary, the failure to support the null hypothesis may actually lead to new information or additional research not previously known or conducted.

Whether it is the research, null, or rival hypothesis, it must be clearly stated and consist of readily identifiable variables. Variables are factors that can change or influence change. They result from the operationalization of a concept. There are two types of variables. Dependent variables are the factors being influenced to change over which the researcher has no controls. The dependent variables are the outcome items or what is being predicted. Independent variables are the factors that influence or predict the outcome of the dependent variable. This variable is something the researcher can control.

Continuing from the Dantzker study, the independent variable is the individual psychologist and associated demographics. The main element is how the respondent labeled himself or herself as a psychologist. This research uses categories that best represent the potential psychologist involved in police psychologic services: Police Psychologist, full time in-house psychologist; Psychologist Consultant, full-time consultant to law enforcement; Clinical Psychologist, occasional service provider to law enforcement; and Other, to allow respondents the opportunity for self-identification into unforeseen categories.

The dependent variables include the types of protocols and the reasons for their use. The protocol choices were garnered from previous research. The choices or reasons were developed from the literature, existing guidelines, and state legislation.

Ultimately, the researcher must identify the variables properly, because misidentification can cause the research to be useless. A better understanding of variables will occur during a later discussion on measurement.

Summary

Before starting any research effort, a topic must be chosen, keeping in mind that the research effort can explore, describe, or explain. This topic should be of interest; be relevant; and have support in the literature through journals, government documents, or the Internet.

On choosing the topic, the research question is created, which advises others what is to be studied. From the research question comes the hypothesis or hypotheses, a statement that indicates the nature of the relationship to be studied. The three main types of hypotheses are the research, null, and rival. The goal is to disprove the null hypothesis.

Finally, for the hypothesis or hypotheses clearly identifiable variables are required. Two types of variables are the dependent, which cannot be controlled by the researcher, and the independent, which can be controlled. Failure to clarify the variables could render the research useless.

METHODOLOGICAL QUERIES

1 Because the topic of the study is already identified, choosing a topic is not necessary. Nonetheless, the sheriff wants to know the issues one should consider before selecting a research topic. What do you tell him?

2 There are three purposes of research. Explain these purposes and how they may apply to the proposed research.

3 Because this will be your first study of job satisfaction, the sheriff wants to know how you will prepare yourself to better understand the concept. Describe to him what a literature review is and the sources that are available for such a review.

4 Give an example of a research question, the types of hypotheses, and possible variables for the proposed study.

The Language of Research

What You Should Know!

Conducting and understanding research may seem to some like a foreign language. There are words, formulas, and so forth that may have more than one meaning or application. For example, the term "research" is often used in more than one context. It is quite common for students and teachers alike to use the term "research" to describe a paper assignment that is actually a literature review. Conducting research comes with a language that one must understand before one can proceed with the research. Therefore, from this chapter the reader should be able to do the following:

1. Define theory and explain how it relates to research.
2. Describe the conceptualization process.
3. Describe what takes place during operationalization.
4. Define variable, dependent, and independent.
5. Describe what a hypothesis is and how it differs from an assumption.
6. Present and discuss the types of hypotheses.
7. Identify a population and discuss how it is related to a sample. Provide examples of some different types of samples.
8. Define validity and reliability. Explain how they connect.
9. Describe what data are. Describe the four levels of data.
10. Discuss the steps in the research process.

The Language of Research

It is quite common for students and teachers alike to use the term "research" to describe a paper assignment that is actually a literature review. As previously noted, with respect to criminal justice and criminology there is more to research than reviewing literature. This synonymous use of the term "research" is just one example of the need to understand associated language. In Chapter 1, the term "research" was defined. In this chapter, various associated terms, such as "theory," "hypothesis," and "variable," are defined or further expanded.

Theory

There is an interesting debate one could have regarding the term "theory," which is reminiscent of the age-old argument: Which came first, the chicken or the egg? With respect to theory, one side of the debate argues that theories drive the research (theory-then-research) or deductive logic. The other side argues that research creates the theory (research-then-theory) or inductive logic (Adler & Clark, 2007; Bryman, 2008; Colton & Covert, 2007; Creswell, 2008).

In reality, as noted in Chapter 1, the two types of logic are actually extensions of one another. Observation may lead to theory construction, which then leads to more observation to test the theory. Therefore, even research that is initially inductive in nature ultimately becomes deductive in that the theory that is generated is tested by observation. In short, all criminal justice practice is grounded in criminologic theory. Theory is defined here as an explanation that offers to classify, organize, explain, predict, or understand the occurrence of specific phenomena.

Based on the definition, a theory is a statement that attempts to make sense of reality. Reality consists of those phenomena that one can identify, recognize, and observe. For example, in criminology, criminal behavior is observed. Therefore, people breaking the law are a reality. A question that arises from this reality is what causes people to break the law? It is here that theory comes into the picture. Criminology is replete with theories about criminal behavior that focus on causes that include biologic, psychologic, and sociologic factors. For examples of theories in criminology, see Box 4-1.

Whether theories have any merit or are truly applicable is why research is conducted. Proving that a theory is valid is a common goal of criminologic and criminal justice researchers. However, to research a theory, the first step is to focus on a concept.

Conceptualization

A concept is best defined as an abstract label that represents an aspect of reality (usually in the form of an object, policy, issue, problem, or phenomena). Every

---

Box 4-1

Examples of Theories in Criminology

Biologic
  A person's physique is correlated to the type of crime one commits.
  Criminality is genetic.
  A chemical imbalance in one's brain can lead to criminal behavior.
Psychologic
  Criminal behavior is the result of an inadequately developed ego.
  Inadequate moral development during childhood leads to criminal behavior.
  Criminals learn their behavior by modeling their behavior after other criminals.
Sociologic
  Socializing with criminals produces criminal behavior.
  Society's labeling of an individual as deviant or criminal breeds criminality.
  Failure to reach societal goals through acceptable means leads to criminality.

---

discipline has its own concepts. For example, in criminal justice and criminology some concepts include criminality, law, rehabilitation, and punishment.

Concepts are viewed as the beginning point for all research endeavors and are often very broad in nature. They are the bases of theories and serve as a means to communicate, introduce, classify, and build thoughts and ideas. To conduct research, the concept must first be taken from its conceptual or theoretical level to an observational level. One must go from the abstract to the concrete before research can occur. This process is often referred to as "conceptualization." As with the definition of theory, there is more than one way to approach conceptualization. This text promotes the two-phase (theory and research levels), five-stage (conceptual level, conceptual components, conceptual definitions, operational definitions, and observational level) approach (Nachmias & Nachmias, 2000). In most research it is seldom specified just how the concept reaches its researchable position. This can cause readers to have problems in understanding what is being researched and why. Therefore, it is often helpful when the researcher can offer readers a clearer picture of the conceptualization of the topic.

To achieve the second part of the conceptualization model, the research phase, the concepts must now be measured. Although concepts can be qualitative, they are most often converted into variables through a process called "operationalization."

Operationalization

The act of operationalizing is the describing of how a concept is measured. This process is best described as the conversion of the abstract idea or notion into a measurable item. It is the taking of something that is conceptual and making it observable, or going from abstract to concrete.

> **FROM THE REAL WORLD**
>
> Seldom do researchers report how they conceptualized their concept. When they do, it provides a better understanding of the research. From the following excerpt, can you fit the pieces into the first phase, theoretical, of the conceptualization model?
>
> > *Community empowerment is a concept used to describe individuals living in close proximity who as a group unite to combat a common problem. The focus of the group is the common problem. If a community is to be empowered, the residents must first be aware that a problem exists (community awareness) to such an extent that it is disturbing or troubling (community concern), resulting in organization of the community (community mobilization) to fight against it (community action). (Moriarty, 1999, p. 17)*

Operationalization is one of the more important tasks before conducting any research. Furthermore, there is no one right way. How this is accomplished is up to the researcher. Yet, it is common for researchers to publish their results without ever explaining how their concepts were operationalized. This shortcoming has made it difficult for many students to comprehend fully the notions of conceptualizing and operationalizing variables. Therefore, when research is conducted that focuses on these two terms, it can be quite useful.

The following excerpt shows how a concept is operationalized. Community awareness was conceptualized as the level of knowledge about the use of alcohol and other drugs in the community. Four variables reflected community awareness: (1) drug usage in the neighborhood; (2) drug dealing in the neighborhood; (3) alcohol and drug prevention messages; and (4) availability of certain drugs (eight different drugs in all). The following are the actual questions used to establish each variable (Moriarty, 1999, p. 18):

- *Drug usage in the neighborhood: Respondents were asked, "How many people in this neighborhood use drugs?" Responses included "many, some, not many or no residents use drugs."*
- *Drug dealing in the neighborhood: Respondents were asked, "How often do you see drug dealing in this neighborhood?" The responses included "very often, sometimes, rarely, never."*
- *Alcohol/drug prevention message: Respondents were asked if they had heard or seen any drug or alcohol prevention messages in the past six months.*
- *Availability of certain drugs: Respondents were asked about the difficulty or ease of obtaining specific drugs in the county. The list of drugs included marijuana, crack cocaine, other forms of*

> *cocaine, heroin, other narcotics (methadone, opium, codeine, paregoric), tranquilizers, barbiturates, amphetamines, and LSD. Each drug availability represents one variable.*

Variables

The primary focus of the operationalization process is the creation of variables and subsequently developing a measurement instrument to assess those variables. Variables are concepts that may be divided into two or more categories or groupings known as "attributes." This ability enables one to study their relationships with other variables. Attributes are the grouping into which variables may be divided. As an example, "male" is an attribute of the variable "gender." There are two types of variables: dependent and independent.

Dependent Variables

A dependent variable is a factor that requires other factors to cause or influence change. They are factors over which the researcher has no controls. The dependent variable is the outcome factor or that which is being predicted. In criminal justice and criminology, criminal behavior is a dependent variable because it requires other factors for it to exist or change. These other factors are the independent variables.

Independent Variables

The independent variable is the influential or the predictor factor. These are the variables believed to cause the change or outcome of the dependent variable and are something the researcher can control. Some better-known independent variables used in criminal justice and criminologic research are gender, race, marital status, and education.

Identifying and recognizing the difference between the variables is important in research, but sometimes may get lost. Therefore, when research specifically calls attention to the variables, it can be quite informative. The key to any research is to be able to operationalize the concepts into understandable and measurable variables. Failing to complete this task makes the creation and testing of the hypotheses more difficult.

Hypotheses

Once the concept has been operationalized into variables fitting the theory in question, most research focuses on testing the validity of statements called hypotheses. The hypothesis is a specific statement describing the expected relationship between the independent and dependent variables. As previously discussed, there are three common types of hypotheses: (1) research, (2) null, and (3) rival.

Research Hypothesis

The foundation of a research project is the research hypothesis. This is a statement of the expected relationship between the dependent and independent variables. The statement may be specified as either a positive (as one increases, the other increases) or as a negative (as one increases, the other decreases) relationship.

Null Hypothesis

Some argue that the results of the research should support the research hypothesis. Others claim that the goal is to disprove the null hypothesis, which is a statement indicating that no relationship exists between the dependent and independent variables. Therefore, a null hypothesis from the previous example might be "organizational variables are not more important than individual variables in predicting an officer's turnover intention."

By rejecting the null hypotheses, the research goal has been fulfilled. However, rejecting the null hypothesis does not necessarily mean that the results have established the validity of the research hypothesis. Validity or significance is discussed later.

Rival Hypothesis

Before starting the research it is customary to establish the research hypothesis, which is generally what the researcher hopes to validate or demonstrate. However, sometimes the results may reject both the null hypothesis and the research hypothesis. This allows for the creation of what is called a "rival hypothesis." The rival hypothesis is a statement offering an alternate prediction for the research findings. Assume that in Lee et al.'s (2009) study the findings were not what they expected or hoped for, instead finding that the difference was opposite. A rival hypothesis might be "participatory climate and internal stress have a direct effect on an officer's turnover intention."

FROM THE REAL WORLD

In examining the effect of participatory management on internal stress, overall job satisfaction, and turnover intention among federal probation officers, Lee, Joo, and Johnson (2009) offered three hypotheses: (1) organizational variables are more important than individual variables in predicting an officer's turnover intention; (2) among organizational variables, participatory climate, internal stress, and overall job satisfaction, respectively, have a significant direct effect on an officer's turnover intention; and (3) participatory climate and internal stress also have a significant indirect effect on an officer's turnover intention.

It is usually the goal of the research to be able to reject the null hypothesis. Testing the research hypothesis becomes central to the research, making identifying the hypothesis an important aspect of the research. Yet, although hypotheses often take center stage in research, there is another type of statement that can find its way into the research: assumptions. However, these types of statements should be avoided whenever possible.

Assumptions

Hypotheses are educated guesses about the relationship between variables, and must be proved by the research. An assumption is a statement accepted as true with little supporting evidence. From a research perspective, assumptions are problematic. It is expected that with statements of inquiry or fact that there be substantiating research. It seems generally inconsistent in scientific research to have assumptions (Banyard & Grayson, 2009; Creswell, 2008).

Assumptions can be defined as statements one accepts as being true with little or no supporting evidence, a stance apparently wholly inappropriate to scientific research (Gillham, 2009). However, it is difficult to conceptualize a piece of research without having some assumptions about the topic of the study. Some may be needed in the way of a working hypothesis or hypotheses to help guide thought about the topic of interest and to help shape the design of the study. It is good research practice to identify, surface, and acknowledge any a priori assumptions of which the researcher is aware while research is in the planning stages.

Fortunately, assumptions can often lead to research. For example, because of the believed natural caring instincts of women, an assumption might be made that women would make better police officers than men. Because there is little evidence to validate this assumption, and it is not a readily accepted statement, at least among men, there is a need to research this assumption. In this situation, the researcher could move beyond the untestable assumption that women would be better officers because they are more caring by converting it into hypotheses that can be tested. Variables could be created to measure what is meant by caring and what is meant by officer performance.

Theory, concept, operationalize, variable, hypothesis, and assumption are all key words in the language of research. Still, they are just the building blocks and causes for other words with which one should be familiar.

Other Necessary Terms

There are many other terms a student should be familiar with before undertaking a research effort. Because these remaining terms are covered in greater detail in later chapters, only a brief definition is offered here, but in the same context as previous definitions.

Once the researcher has managed to conceptualize and operationalize his or her research, it is then time to choose who will be targeted to respond to the dependent variables. A unit of analysis is the level at which the researcher will focus his or her attention. It could be individuals, groups, communities, or even entire societies depending on the nature of the research. The researcher first identifies the target population to be studied, from which then a sample can be drawn.

Population

A population is the complete group or class from which information is to be gathered. For example, police officers, probation officers, and correctional officers are each a population. Although it would be great if every member of a population could provide the information sought, it is usually logistically impractical in that it is both inefficient and wasteful of the researcher's time and resources. Therefore, most researchers choose to obtain a sample from the targeted population.

Sample

A sample is a group chosen from within a target population to provide information sought. Choosing this group is referred to as "sampling," and may take one of several forms. Sampling is important enough to warrant an entire chapter of its own later in the text. Some examples of samples follow.

Random: A random sample is one in which all members of a given population had the same chances of being selected. Furthermore, the selection of each member must be independent from the selection of any other members.

Stratified Random: This is a sample that has been chosen from a population that has been divided into subgroups called "strata." The sample is comprised equally of members representing each stratum.

Cluster: The sample is comprised of randomly selected groups, rather than individuals.

Snowball: This sample begins with a person or persons who provide names of other persons for the sample.

Purposive: Individuals are chosen to provide information based on the researcher's belief that they will provide the necessary information. This type of sample is also known as a "judgmental" or "convenience sample."

Once the sample has been identified, the information is collected. The various collection techniques are covered in detail in a later chapter. In collecting this information two concerns for the researcher are the validity and the reliability of the data collection device.

Validity

Validity is a term describing whether the measure used accurately represents the concept it is meant to measure. There are four types of validity: (1) face,

(2) content, (3) construct, and (4) criterion. There are some individuals who also offer validity as internal (truthfulness of the findings with respect to the individuals in the sample) and external (truthfulness of the findings with respect to individuals not in the sample).

Face validity is the simplest form of validity and refers to whether the measuring device seems, on its face, to measure what the researcher wants to measure. This is primarily a judgmental decision. In content validity each item of the measuring device is examined to determine whether the element measures the concept in question. Construct validity inquires as to whether the measuring device does indeed measure what it has been designed to measure. It refers to the fit between theoretical and operational definitions of the concept. Criterion (or pragmatic) validity represents the degree to which the measure relates to external criterion. It can either be concurrent (does the measure enhance the ability to assess the current characteristics of the concept under study?) or predictive (the ability accurately to foretell future events or conditions).

Reliability

Reliability refers to how consistent the measuring device would be over time. If the study is replicated, will the measuring device provide consistent results? The two key components of reliability are stability and consistency. Stability means the ability to retain accuracy and resist change. Consistency is the ability to yield similar results when replicated.

Having established the validity and reliability of the measuring device, the sample can now be approached for information. The information gathered is known as "data."

Data

Data are simply pieces of information gathered from the sample that describe events, beliefs, characteristics, people, or other phenomena. These data may exist at one of four levels: (1) nominal, (2) ordinal, (3) interval, and (4) ratio.

Nominal data are categorical based on some defined characteristic. The categories are exclusive and have no logical order. For example, gender is a nominal level data form. Ordinal data are categorical too, but their characteristics may be rank-ordered. These data categories are also exclusive but are scaled in a manner representative of the amount of characteristics in question, along some dimension. For example, types of prisons may be broken down into the categories of minimum, medium, and maximum. Categorical data for which there is a distinctive, yet equal, difference among the characteristics measured are interval data. The categories have order and represent equal units on a scale with no set zero starting point (e.g., the IQ of prisoners). Ratio data are ordered; have equal units of distance; and a true zero starting point (e.g., age, weight, or income).

As the text continues, other terms are introduced and defined. Because a sufficient number of terms have been introduced, it is now possible to review the research process in a research manner.

The Research Process

Having been introduced to research and its language, the last item to be offered in this chapter is a model of the research process through terminology. This model begins with a theory usually identifying some concept. The concept is then conceptualized and operationalized creating dependent variables. Completing the identification of both the independent and dependent variables leads then to developing the hypothesis. Finally, a sample is chosen, measurement or information is gathered from the sample, the information is converted into the proper data for analysis, and the results are reported. This process becomes functionally clearer as the text progresses.

Summary

Becoming proficient in research requires knowing the language. Several terms have been introduced that are important to mastering research as a language. The main terms include theory, concept, operationalize, variables, hypothesis, and sample. There are two types of variables: independent and dependent. A sample may be random, stratified, clustered, snowball, or purposive. Other terms are validity (face, content, construct, and criterion); reliability; and data (nominal, ordinal, interval, and ratio). With knowledge of these terms, the research process can be taken to another level.

METHODOLOGICAL QUERIES

1 Job satisfaction is the main concept for the proposed study. Describe this concept through the conceptualization process.
2 To conduct the study, you'll need to operationalize job satisfaction. Describe what takes place during operationalization and how it would work for job satisfaction.
3 Identify at least one dependent and one independent variable.
4 Identify the population for the proposed study and explain why it may require sampling.
5 The sheriff wants to understand how validity and reliability are important to the research process. Explain their application.
6 What type of data might you collect in the proposed study? Describe the four levels of data and how they might each be applied to the study.

Qualitative Research

5

What You Should Know!

A long-standing debate among academic researchers is over what type of research is more acceptable, qualitative or quantitative. The debate about which form of research is more appropriate in criminal justice tends to pit theorists against applied practitioners. Interestingly, as the debate continues, one could argue that both have a place in criminal justice. Therefore, students should have enough knowledge to differentiate between the two types of research. From this chapter the reader should be able to do the following:

1. Compare and contrast quantitative and qualitative research.
2. Define qualitative research.
3. Discuss the strengths and weaknesses of qualitative research.
4. Explain how qualitative research differs from a literature review.
5. Describe the types of field interviews and give examples.
6. Describe the different roles that may be used in field observation.
7. Describe ethnographic research.
8. Describe sociometry.
9. Describe historiography.
10. Describe how a content analysis may be qualitative.

Because of its long-standing image of being an applied social science, some of the criminal justice research conducted and published has had detractors. This is primarily the result of what is perceived as the research's lack of statistical sophistication. This perception is central to the continuing debate over what is more "academic," qualitative or quantitative research. Because the remainder of the text focuses on elements more commonly associated with quantitative research (although very applicable to qualitative research) this chapter offers a closer look at qualitative research.

Qualitative Versus Quantitative Research

In Chapter 1, it was briefly noted that the debate over qualitative versus quantitative research simply comes down to a question of concepts as ideas or terms versus numerical values. Broadening this distinction in easy terms offers quantitative research as merely referring to counting and measuring items associated with the phenomena in question. Qualitative research involves examining a topic through a concept or a symbolic description of factors (Given, 2008). Neither statement tends to clarify why one form of research is better than the other.

In recent years there has been a trend among scholarly journals to demand more quantitative than qualitative research. This preference is because qualitative research is often criticized as not being scientific (Bergman, 2009; Creswell, 2008; Given, 2008). That debate will not be continued here because both methods are appropriate and necessary to criminal justice and criminology.

Qualitative Research Defined

Qualitative research is defined as a nonnumerical explanation of one's examination and interpretation of observations the purpose of which is to identify meanings and patterns of relationships (Creswell, 2008; Hagan, 2006; Maxfield & Babbie, 2009). This type of research encompasses interpreting action or meanings through a researcher's own words (Adler & Clark, 2007) rather than through numerical assignments. Such analysis enables researchers to verbalize insights that quantifying of data does not permit. It also allows one to avoid the trap of false precision, which frequently occurs when subjective numerical assignments are made. These quantifications are misleading in that they convey the impression of a precision that does not really exist.

Merits and Limitations of Qualitative Research

The insights gained from qualitative research and its usefulness in designing specific questions and analyses for individuals and groups make it invaluable in the study of criminal justice and criminology. However, the costs and time involved in such studies may not be logistically feasible (Bergman, 2009; Drake & Jonson-Reid, 2008). One of the major complaints about qualitative research is that it takes too long to complete. Other complaints about qualitative research include that it requires clearer goals and cannot be statistically analyzed (Creswell, 2008; Given, 2008; Maxfield & Babbie, 2009). There may also be problems with reliability in that replication may prove quite difficult. Lastly, validity issues may arise from the inability to quantify the data.

What You Have Done Before

Before we proceed further let us deal with an issue that may be developing in your mind as you read about qualitative research. We have stressed that most, if not all, of the research that you have done in the past was not true research but merely an extended literature review. Now we are telling you that research does not have to use numbers. Therefore, we were obviously wrong and you have been doing research all along. This is a false assumption.

Unless one has used the scientific method to structure an inquiry so as to yield results that logically extend from the analysis, one has not conducted qualitative research. A mere reciting of what others have done does not qualify as research. However, if one actually used the prior research to compare or assess an issue or event one may actually have conducted qualitative research.

The distinction here between true qualitative research and the literature review "research" becomes clearer after one reviews the various types of qualitative research that are available for use. There are a variety of methods available for conducting qualitative research. They include field interviews, focus groups, field observation, ethnography, sociometry, and historiography (Creswell, 2008; Given, 2008; Maxfield & Babbie, 2009).

Types of Qualitative Research

Conducting qualitative research may be time-consuming, but because it better reflects the actual being of something, the time factor becomes moot for the qualitative researcher. Therefore, the bigger issue becomes what method to use. This depends on the goals of the research and from where the information is to be sought. When information is wanted directly from individuals, the most common method to use is interviewing.

Field Interviewing

Interviewing is the asking of questions by one individual of another to obtain information. If the interview consists of specific questions for which designated responses may be chosen, this qualifies as quantitative research (quantitative interviews are discussed in the next chapter). Generally, even if a field interview is structured the answers are open-ended. By this we mean that the response given by the interviewee is recorded exactly as stated rather than assigned to a predetermined category.

There is no one agreed way to conduct a field interview. Consensus could probably be found that a major key to interviewing is asking the right questions. A reflection on the questions would be based on the type of interview being conducted and the information sought. Although there may be debate as to what the differing ways of interviewing are called, three types of interviews are possible: (1) structured, (2) semistructured, and (3) unstructured (Dunn, 2009).

Structured Interviews

A structured interview entails the asking of preestablished open-ended questions of every respondent. As stated previously, if the questions asked are closed-ended (respondents select from a choice of predetermined answers for which a numerical value can be assigned), this qualifies as quantitative survey research. Most structured interviews are quantitative in that they consist entirely or predominantly of closed-ended questions. Responses are recorded as given and the interview pace is such that all the questions can be asked and responded to in a timely fashion, but neither the interviewer nor the respondent are rushed. The following have been suggested as guidelines for conducting a structured interview. Never get involved in long explanations of the research, deviate from the study introduction, sequence of questions, or question wording, let another person interrupt the interview, or respond for the person being questioned, suggest, agree to, or disagree to, an answer, interpret the meaning of a question, or improvise.

The structured interview is geared toward limiting errors and ensuring a consistency of order in the responses even though the responses themselves may vary. Still, there are factors that could cause problems eventually affecting the outcome. Some errors include, but are not limited to, the respondent's behavior (e.g., is the respondent being truthful or only saying what he or she believes the interviewer wants to hear); the setting of the interview (e.g., face-to-face or telephonic); question wording (e.g., uncommon terms); and poor interviewer skills (e.g., the interviewer does not clearly enunciate or changes the question's wording). In addition, if an interviewer is not familiar with the

---

FROM THE REAL WORLD

Dantzker and McCoy (2006) used a structured interview to obtain information regarding psychologic screening of police candidates from the 17 largest municipal police agencies and the Department of Public Safety in Texas. Once contact was made with the appropriate person, two basic questions of a structured interview were asked and responses recorded: What psychologic protocol(s) are used to screen police applicants, and what is the general content of the psychologic interview?

FROM THE REAL WORLD

Charles (2009) examined spirituality and law enforcement using semistructured interviewing, designed to explore the expression of spirituality as revealed by the officers in their work. This particular method of narrative inquiry allowed the participants to "tell their stories." The author asked the officers eight standardized, open-ended questions.

1. When did you become a law enforcement officer and why?
2. Tell me your spiritual history starting with your parents.
3. Describe your spiritual practice.
4. Tell me about your work as an officer.
5. Has your spirituality influenced your work as an officer?
6. What has been most challenging to you while working in a profession where you are constantly exposed to human destructiveness and suffering?
7. How have you changed as an officer? How has your spirituality helped?
8. How do you cope with the human destructiveness and suffering encountered in police work? What is your support system when you are overwhelmed?

respondent's background, culture, education, or other factors, this can be detrimental to the interview.

The structured interview can elicit rational, legitimate responses. Unfortunately, it does not consider the emotional aspect. This is where a semistructured interview might be more useful.

Semistructured Interviews

This type of interview primarily follows the same ideas or guidelines of a structured interview. The major difference is that in this type of interview, the interviewer can go beyond the responses for a broader understanding of the answers. This is known as "probing for more detail." Probing may consist of asking for more explanation of an answer than has been given or following-up with an additional question or questions depending on the answers given. Semistructured interviews are commonly used as a qualitative research strategy.

Unstructured Interviews

The unstructured interview is far less rigid than either of the previous two methods. Seldom is a schedule kept or are there usually any predetermined possible answers. Often the questions are created as the interaction proceeds (Adler & Clark, 2007). The most common form of unstructured interview makes uses of open-ended (ethnographic or in-depth) questions. In many cases, this style interview is done in conjunction with participant-observation. For example, to gain a better understanding of what police detectives find stressful, a researcher chose to spend several days with detectives from a particular police agency. Although the primary role was to observe how detectives responded to certain situations or events, during "down times" (e.g., between calls, meals, after the shift) the researcher was able to ask open-ended questions pertaining to what the detectives found stressful. One such question might be "What do you find stressful about being a detective?"

Because of the nature of an unstructured interview, the researcher must be able to complete the following for the study to be successful (Adler & Clark, 2007; Creswell, 2008; Dunn, 2009; Given, 2008; Maxfield & Babbie, 2009):

1. Gain access to the setting
2. Understand the language and culture of respondents
3. Decide how to present oneself
4. Locate a contact or informant
5. Gain the respondent's trust
6. Establish rapport

By meeting these requirements, the researcher should be successful in his or her efforts.

Interviewing can be a tedious means of gaining data, especially if it is of the one-on-one variety. However, sometimes research requires interviewing more than one person at a time. This qualitative method is often called "focus groups" or "group interviewing" (Dunn, 2009; Given, 2008).

Focus Groups

Perhaps the best way to define a focus group is the interviewing of several individuals in one setting. Although not meant to replace individual interviews, focus groups have long been in use in marketing and politics (Given, 2008). The focus group is an information gathering method where the researcher-interviewer directs the interaction and inquiry. This can occur in either a structured (e.g., pretesting a questionnaire) or unstructured (e.g., brainstorming) manner. In either case the researcher–interviewer must meet the same guidelines offered for conducting any interview. Why, however, would someone use this method?

The use of focus groups has its advantages and disadvantages. The advantages include limited expenses, flexibility, and stimulation. The disadvantages include group culture, dominant responder, and topic sensitivity (Adler & Clark, 2007; Creswell, 2008; Dunn, 2009; Given, 2008; Maxfield & Babbie, 2009). The focus group can be a useful qualitative method for gathering information and can be particularly interesting. However, there may be times when the researcher may wish to take part in the activities or just observe the subjects of the research in their natural setting, a method of research referred to as "observational."

Field Observation (Participant–Observer)

The method of observation is interesting because of the context in which it places the researcher. Although this method of qualitative research has not received the same attention as interviewing (Adler & Clark, 2007; Dunn, 2009), it is still a very viable tool especially in criminal justice and criminology. Usually this method of study has only focused on two means: observer and participant–observer.

The observer method is where the researcher gathers information in the most unobtrusive fashion by simply watching the study subjects interact, preferably without their knowledge. With participant–observation, the subjects not only are aware of the researcher but actually interact as the researcher takes an active role in activities under study. Although these two methods are quite fitting, a more useful description of the options for this type of research had been offered over a decade ago. This model includes four types: (1) full participant, (2) participant researcher, (3) researcher who participates, and (4) the complete researcher (Senese, 1997).

The Full Participant

Sometimes it may be fitting or perhaps even essential that a researcher become part of the study. The full participant method allows the researcher to carry out observational research, but does so in a "covert" manner (e.g., to study the workings of a gang, the researcher takes an active role as a gang member, where other members are unaware of the research agenda). Some problems with this method include the researcher possibly facing ethical and moral dilemmas that could adversely affect the research if the wrong decision is made, such as having to compromise one's beliefs, or even place the person in legal jeopardy (e.g., being in a car during a drive-by shooting). To avoid ethical and moral dilemmas (discussed in Chapter 2), a researcher may decide to move one step down, and become a participant researcher.

The Participant Researcher

Because participating in the activities of the research subjects can offer insights and information not attainable through other forms of research, and to avoid dilemmas facing full participation, there is the participant research method. This method is where the researcher participates in the activities of the research environment but is known to the research subjects to be a researcher (e.g., to study the behavior of males in a reformatory, the researcher goes in and participates in all activities, but everyone knows he or she is there collecting information). The biggest negative of this research method is that by the subject knowing the researcher's role, behaviors other than those being studied can be influenced. The subjects may not act as they would under unobserved conditions, either overexaggerating or underexaggerating their actions.

This is known as "subject reactivity" to the research. Furthermore, by being part of the activities, the researcher can influence outcomes and behaviors that may not have existed without his or her presence. This then gives way to the third method of observation.

The Researcher Who Participates

Rather than taking part in activities, the "researcher who participates" method requires nothing more than observation by the researcher whose status as a researcher is known to the research subjects. For example, to study whether a particular treatment offered to incarcerated juveniles is working, the researcher enters the environment where his or her status as a researcher is known, but does nothing more than observe. Although this method eliminates many of the problems of the previous two methods, the mere presence of the researcher who participates can still influence behavior and activities. How does one avoid this dilemma?

The Complete Researcher

The way for a researcher to minimize the problems generally associated with observation–participatory research is to avoid all possible interaction with the research subjects. Data collection may involve covert methods of observation (e.g., from some disguised vantage point, such as a guard tower in a maximum level prison or the review of records). As can be imagined, the benefits that are gained by covert observation or assumption of a noninteracting role are countered by possible resentment and denial of access by those being observed.

Regardless of which form is used, observational research can be time-consuming. It may not provide the results ultimately sought. Still, it can be a very interesting means for gathering data.

FROM THE REAL WORLD

To study the implementation of community policing, Memory (1999) conducted research as the "researcher as participant." His methodology took the form of ride-alongs, where he spent 180 hours riding with police officers from selected cities or counties. Departmental contact persons arranged the ride-alongs ahead of time. Because six law enforcement agencies had agreed to be involved in a larger overall research project, the researcher did not need to "gain access," but still had to comply fully with agency and officer directions to maintain access. Memory rode during both daylight and darkness in each jurisdiction but rode nearly none between 1:00 AM and 1:00 PM. As a researcher as participant, he dressed in slacks, a dress shirt, and a sport coat. He observed everything that occurred concerning the officers with whom he rode. The data collection involved taking notes on the front and back of data-collection forms. The researcher recorded observations, words of officers and citizens, and his own impressions and ideas.

Despite the respective problems of observational research it does provide a perspective that is often missed through quantitative research. Furthermore, it can be an important part of another form of qualitative research: ethnography.

Ethnographic Study

Ethnographic study or field research (ethnography) overlaps with field observation in that the researcher actually enters the environment under study, but does not necessarily take part in any activities. The defining of ethnographic study has been inconsistent but may best be viewed as a form of qualitative research in which the researcher examines real social structures for various sociologic, psychologic, and educational variables (Berg, 2008; Creswell, 2008). It tends to consist of several attributes that include exploring the nature of particular social phenomena, a tendency to work primarily with unstructured data, the investigation of a small number of cases, and analyzing data involving explicit interpretation of the meanings and functions of human actions (Berg, 2008; Creswell, 2008). The product of such a study is verbal descriptions and explanations where quantification and statistical analysis are of minimal support (Given, 2008).

Not a particularly popular means of conducting research, ethnography, like observation, can provide insights not found through quantitative research. For example, the study of gang graffiti from a quantitative perspective might simply identify the differing types of symbols drawn and how many there are of each. However, an ethnographic study could provide what the symbols mean, why they have been placed where, and who is placing them. This information gives a different slant to the information than the quantitative approach. One of the interesting aspects of ethnographic study is the possibility for the researcher to examine social interactions, which leads to another form of qualitative research: sociometry.

Sociometry

Sociometry is a technique by which the researcher measures social dynamics or relational structures within a specific environment (Berg, 2008; Creswell, 2008; Givens, 2008). Information can be gathered through interviews or by observation and indicates who is chosen and the characteristics about those who do the choosing. With respect to criminal justice, a sociometric study might involve prosecutor–defense attorney relationships, where the researcher observes the interaction of prosecutors with defense attorneys noting how each treats and is treated by the other and how that affects outcomes, such as plea bargains. This study might show whether there is a hierarchy among lawyers or among potential defendants (the accused).

As with observation and ethnography, the researcher's presence, attitudes, biases, and so forth can influence the outcomes of sociometric research.

FROM THE REAL WORLD

Harpster, Adams, and Jarvis (2009) examined verbal indicators to analyze 911 homicide statements critically for predictive value in determining the caller's innocence or guilt regarding the offense. They listened to 100 audio recordings and transcripts of 911 homicide telephone calls obtained from law enforcement departments throughout the United States formulating the linguistic attributes of these communications and identifying a variety of variables that they then analyzed for association with the likelihood of the caller's guilt or innocence regarding the offense of homicide.

Because the mere presence of the researcher can be problematic, a less obtrusive means of conducting qualitative research is available.

Historiography

Historiography is the study of actions, events, and phenomena that have already occurred (Berg, 2008; Creswell, 2008; Givens, 2008). This type of research often involves the study of documents and records with information about the topic under study. This type of study is generally inexpensive and unobtrusive. It can assist in determining why or how an event occurred and whether such an event could happen again. It is also a means by which researchers may compare and contrast events or phenomenon that have occurred. Historiography can be either qualitative or quantitative in nature depending on the materials being used and the focus of the research. This method is used by historians and Marxist scholars.

Content Analysis

Like historical research, content analysis is the study of social artifacts to gain insights about an event or phenomenon. It differs in that the focus is on the coverage of the event by the particular medium being evaluated (books, magazines, television programs, news coverage, and so forth) rather than the event. Depending on how the research is conducted, this may be either qualitative or quantitative in nature. Qualitative content analysis emphasizes verbal rather than statistical analysis of various forms of communication. For example, Klinger and Brunson (2009) conducted a content analysis of detailed accounts by police officers of how they perceived what transpired during incidents where they shot citizens.

Overall, qualitative research offers a perspective that contrasts, yet can compare to that which is discovered through quantitative research. Perhaps Berg (2008) best summarizes qualitative research:

Qualitative research properly seeks answers to questions by examining various social settings and the individuals who inhabit these settings. Qualitative researchers, then, are most interested in how humans arrange themselves and their settings and how inhabitants of these settings make sense of their surroundings through symbols, rituals, social structures, social roles, and so forth (p. 7).

Whether it is qualitative or quantitative, the research can serve a purpose in criminal justice and criminology. Thus, the debate should not be over which is better, but over what should be studied. How the selection of research impacts on the research design is discussed in a later chapter.

Summary

There is a continuing debate between criminal justice and criminology researchers as to what type of research is best, qualitative or quantitative. This chapter argues that they complement each other and have appropriate roles in related research. To that end, this chapter explores what qualitative research is by noting that it is the research of ideas and concepts, not of numbers. Although there are many forms of qualitative research, the more popular include the following:

1. Field interviewing: structured, semistructured, and unstructured
2. Field observation: the full participant, participant researcher, the researcher who participates, and the complete researcher
3. Ethnography
4. Sociometry
5. Historiography
6. Content analysis

METHODOLOGICAL QUERIES

1   Since a decision has not been made as to how the proposed study will be conducted, the sheriff is interested in whether you would observe behaviors or conduct a survey. Explain the difference between quantitative and qualitative research to him.

2   The sheriff thinks that simply observing the officers to determine their satisfaction would be adequate. How would you discuss the strengths and weaknesses of qualitative research?

3   You explain to the sheriff that observing behaviors or field observation is a feasible way to conduct a study. How would you describe to him the different roles that may be utilized in field observation?

4   Part of field observation could involve field interviews. You must explain to the sheriff the different types of field interviewing. Be sure to demonstrate how that might apply to the proposed study.

CHAPTER

6

Quantitative Research

What You Should Know!

As noted in Chapter 5, there is a long-standing debate among academic researchers over what type of research is more acceptable, qualitative or quantitative. Chapter 5 introduced qualitative research. This chapter examines quantitative research, that is, how statistics are used for completing research. From this chapter the reader should be able to do the following:

1. Define quantitative research.
2. Define empiricism and discuss how it relates to quantitative research.
3. Contrast the two types of causality.
4. Identify and discuss the three criteria for causality.
5. Contrast necessary and sufficient cause.
6. Describe what is meant by false precision.
7. Identify and explain the four levels of measurement.
8. Recognize and describe the types of survey research.
9. Explain the strengths and weaknesses of survey research.
10. Describe field research and give an example.
11. Discuss the types of unobtrusive research.
12. Describe evaluation research.

As stressed in Chapter 5, qualitative research is a valuable tool that provides many insights into the study of crime, criminality, and society's responses to crime. It is not only an excellent means of conducting primary research but is highly useful in complementing quantitative research.

However, to be a complete criminologic researcher, one must be able to move beyond qualitative research. Quantitative research is the means to do so.

Quantitative Research

Unlike qualitative research, the definition of which has been controversial and often inconsistent, definitions of quantitative research are quite consistent: it provides a means of describing and explaining a phenomenon through a numerical system (Berg, 2008; Fowler, 2009; Maxfield & Babbie, 2009). This means that research is not based on a possibly subjective interpretation of the observations but is a more objective analysis based on the numerical findings produced from observations.

The previous chapter noted that quantitative research refers to counting and measuring items associated with the phenomena in question, whereas qualitative research focuses on concepts and verbal description. Because of the potential for bias and the criticisms of qualitative research as being unscientific, most criminologic research tends to be quantitative in nature. A quick review of the leading journals that publish criminal justice and criminology research supports this assertion.

There are many issues that are not suitable for numerical assignments; to do so would be inaccurate and misleading. Some of the more intense theoretical debates, such as the merits of the death penalty, are based on personal beliefs about human nature that are shaped by deep-seated religious, political, and moral convictions. As a result, perceptions of these tend frequently to be more influenced by emotion and ideology than by scientific study. This does not mean that quantitative research cannot be conducted, but it is difficult to do so.

What You Have Not Done Before

Developing researchers may believe that they have done qualitative research through the literature reviews and comparative and historical research done for previous courses. If done in a systematic and logical manner consistent with the scientific method, perhaps this is true. But have numerical values been assigned, data collected using that assignment, and the results analyzed? It is doubtful, unless one has a strong background in math and science or has been fortunate enough to have been involved in an empirical research project (for those who are still struggling with methods phobia, fortunate is defined here as having benefited from the knowledge and insights of the experience rather than from the pleasure of the experience). This is a deficiency that the authors hope to aid in correcting. A college graduate who values and is capable of critical thinking and independent learning needs to be able to conduct quantitative research. It will prove valuable not only in an academic career but in future work tasks or civic duties. It is surprising to find how many things there are in life that warrant "looking into." In addition, quantitative research is actually easier than qualitative research.

Empirical Observation

Thinking back to an introductory course in criminology and criminal justice or a course in criminologic theory, one will realize they have already received an introduction to empiricism. Cesare Lombroso, the founder of the Positive School of Criminology, used empiricism in his study of criminals. Earlier scholars (e.g., Quetelet and Guerry in their independent area studies of crime rates in France) also used empirical techniques.

Empiricism is use of sensations and experiences (observations) to arrive at conclusions about the world in which we live (Jeffery, 1990). The use of the scientific method with its focus on causation rather than casual observation is what makes empiricism important. It is this emphasis on empiricism rather than idealism that is the basis on which positive criminology is founded. Rather than just quoting an "eye for an eye," empirical techniques can be used to gather and evaluate data as to the effectiveness of correctional programs, or to decide which patrol strategy is more cost effective, or to determine whether extralegal factors influence conviction rates. Quantitative research is based on empiricism.

Causality

In applying empirical observation to criminal justice research the focus is on causal relationships. Simply stated, what behaviors or events lead to other behaviors or events? When trying to answer that question, one seeks to determine causality.

Idiographic and Nomothetic Causes

The examination of numerous explanations for why an event occurred is known as "idiographic explanation." Historians and Marxist criminologists tend to use this method in their qualitative analyses. Idiographic causation may also be quantitative in that many causes may be compared and contrasted using numerical assignments. When researchers focus on a relatively few observations to provide a partial explanation for an event, this is known as "nomothetic causation" (Berg, 2008; Bergman, 2009; Gavin, 2008).

Rather than trying to provide a total picture of every influence in a causal relationship, nomothetic explanation focuses on one or a few factors that could provide a general understanding of the phenomena being studied. Nomothetic explanations of causality are based on probabilities (discussed in Chapter 9). It is this use of probability that enables one to make inferences based on a relatively few observations.

The Criteria for Causality

In investigating to determine if there is a causal relationship between the events or issues that being studied, three criteria must be observed (Adler & Clark, 2007; Dunn, 2009): (1) the independent variable (the variable that is providing the influence) must occur before the dependent variable (the variable that is being acted on); (2) a relationship between the independent and dependent variable must be observed; and (3) the apparent relationship is not explained by a third variable.

For example, one sees individual A struck by person B. Person A then falls down. Using the criteria for causality one would conclude that the striking (independent variable) led to the falling (dependent variable). It occurred before the falling and was clearly related to that which was witnessed. If no other event (a third variable, such as a blow having also been struck by a Person C) occurred, then it is reasonable to assume that the criteria for causality have been met.

Necessary and Sufficient Cause

In investigating causality one must meet the previously mentioned criteria, but one does not have to demonstrate a perfect correlation. In probabilistic models, such as those used in most inferential research, one often finds exceptions to the rule. If a condition or event must occur for another event to take place, that is known as a "necessary cause." For example, to collect a paycheck, one must be employed. The cause must be present for the effect to occur.

When the presence of a condition ordinarily causes the effect to occur, this is known as a "sufficient cause." The cause usually, but not always, creates the effect. Playing golf in a thunderstorm may not result in being struck by lightning, but the conditions are sufficient for it to occur. In social science research it is preferred to identify a necessary cause for an event but to do so is often impossible. Instead, one is more likely to identify causes that are sufficient.

False Precision

When quantifying data it is imperative that the numerical assignments be valid. If one arbitrarily assigns numbers to variables without a logical reason for doing so, the numbers have no true meaning. This assignment is known as "false precision." One has quantified a concept but that assignment is subjective rather than objective. The precision claimed really does not exist. In those cases a qualitative analysis is more appropriate.

Quantitative Measurement

On many occasions people are confronted with some form of measurement. Based on the previously mentioned definition of quantitative research, measurement, as it is applied here to criminal justice and criminologic research, is viewed as the assignment of numerical values or categorical labels to phenomena for the expressed purpose of quantifiable identification or analysis. With respect to quantifying something, there is more than one means or level.

Levels of Measurement

Determining what level of measurement to use can often be a very confusing part of conducting quantitative research. There are four levels of measurement from which to choose: (1) nominal, (2) ordinal, (3) interval, and (4) ratio. The level that is chosen has an extremely important impact on how the data are collected and analyzed. The researcher can move down from a higher level of data to a lower level during data analysis but they cannot move up to a higher level from a lower level. The reason for this becomes clear as the various levels of measurement are discussed.

Nominal Level Data

The simplest level of measurement is the nominal level. At this level, measurement is categorical where each is mutually exclusive. There is neither a quantitative nor statistical value assigned except for the expressed need to describe the results or to code it for data analyses. Most often nominal measures are used to represent independent variables that describe characteristics of the sample, but may also be used to describe dependent variables.

Ordinal Level Data

The second level of measurement is the ordinal level. This level moves beyond being merely categorical by assigning a rank or a placement of order to variables. Although in using this level numbers are assigned for ranking purposes (e.g., 1–10), these numbers are not meant to or are unable to explain the response, but are simply viewed as a demonstration of where the respondent believes the item to fall. For example, in looking at the difference between the seriousness of criminal offenses where murder is labeled as a nine, robbery a five, and theft a one, there is a four-unit difference between each but one cannot explain what that difference truly represents.

Most often ordinal measures are found in attitudinal surveys, perceptual surveys, quality of life studies, or service studies. For example, individuals could be given a list of occupations that respondents could be asked to rank in order of how stressful they perceive each to be, from most stressful to least stressful. Although they would write any number between 1 and 10, the listed order would only tell what the individuals perceive. There is no way to determine how much difference there is in the perceived stressfulness between each occupation. Despite being assigned numerical values and being useful in data analysis, ordinal level data are limited in providing explanation.

Interval Level Data

The third highest level of measure is interval, which provides far better opportunity for explanation than data collected at the nominal and ordinal levels. Within interval data there is an expected equality in the distance between choices on the continuum. This allows one to use more sophisticated techniques during data analysis.

Unlike ratio level data (discussed next), there is no set zero or starting point for interval data. Because numbers assigned have an arbitrary beginning, the usefulness of the information may be limited. For example, the difference between an IQ of 135 and 150 is the same 15-unit difference as between 150 and 165. However, there is no distinction as to what the difference means. One can comment on the differences but cannot explain what that difference means. To be able to do so requires ratio level data.

Ratio Level Data

The highest and most quantifiable level of measure is ratio. This level is primarily characterized by an absolute beginning point of zero and the differences between each point are equal and can be explained. Two of the more common ratio measures are age and money. With respect to research, ratio measures can be collapsed into nominal or ordinal measures. Age and family income are often ratio level, independent variables that for analysis purposes are collapsed into a nominal measure (e.g., under 18, 18–25, 26–34, over 34; and under $10,000, $10,000–$24,999, $25,000–$49,999, over $50,000). The benefits of doing so are discussed in a later chapter.

Choosing a level of measurement is just one of the first steps in conducting quantitative research. Probably the most important step is deciding what type of method or design to use.

Types of Quantitative Research

There are a variety of designs available to the justician and criminologist for conducting quantitative research. Yet, despite all the possibilities, there is one form of research that is both a research method and a research tool: the survey. Because of its dual nature, it is discussed briefly here as a quantitative research design and later in Chapter 9 as a data collection device.

Survey Research

The survey is one of the most popular research methods in criminal justice. The survey design is used when researchers are interested in the experiences, attitudes, perceptions, or beliefs of individuals, or when trying to determine the extent of a policy, procedure, or action among a specific group. Most often a researcher contacts a sample of individuals presumed to have participated in a particular event, who belong to a certain group, or who are part of a specific audience having experienced similar events. Of the identified group, certain questions pertaining to the topic under study are asked, either verbally or in written form. The solicited responses to the questions comprise the data used to test the research hypothesis. The three primary survey methods are (1) personal interview, (2) mail questionnaire, and (3) telephone survey.

Personal Interviews

Personal interviews are surveys administered by face-to-face discussions between the researcher and the survey respondent. Unstructured interviews in which the researcher's questions are developed during the conversation are qualitative and were discussed in Chapter 5. Structured interviews in which the researcher asks only open-ended questions also are qualitative.

Interviews in which the researcher reads from a previously developed questionnaire to which responses are numerically assigned are quantitative. Such interviews may in fact be nothing more than the reading of a questionnaire that could have been mailed out to respondents. Personal interviews permit the researcher to obtain not only responses to the questions asked, but permit probing on open-ended questions and observation as to the respondents' demeanor and nonverbal responses to the questions asked. They are, however, more costly and time consuming and less safe than other strategies that do not bring researchers into physical contact with respondents.

Mail Questionnaires

Mail questionnaires are survey instruments mailed to selected respondents to complete on their own rather than being directly interviewed by the researcher. They are much cheaper and safer to administer. In addition, they enable researchers easily to survey large numbers of people very quickly.

Mail questionnaires may be administered three ways. The questionnaire may be mailed with a request for it to be completed and returned in a self-addressed stamped (or metered) envelope to the researcher. To provide a more personal touch the researcher may drop off the questionnaire in a face-to-face contact with the respondent with the request that it be mailed back to them. Alternatively, in a strategy designed to intimidate people into completing the questionnaire (which is not recommended) the researcher may mail out the questionnaire and advise that he or she will come by at a later time to retrieve it. Either of the latter two strategies greatly increases the time and cost involved in conducting the survey.

Telephone Surveys

The last form of survey has rapidly become popular among pollsters. Telephone surveys are quick and easy to do, enabling the researcher to contact large numbers of people in an efficient manner. They are safe in that verbal abuse is about the worst that researchers can incur from their dealings with respondents. Telephone surveys may be more efficient in that they may use random digit dialing to sample the population. They may also have the added advantage of actually inputting data into a computer as the surveyor asks the questions. The biggest disadvantage is that people who are sick of telephone solicitors often refuse to participate. The means of administering survey research are discussed in more detail in Chapter 10.

Pros and Cons of Survey Research

There are numerous reasons why the survey design is popular: (1) it uses a carefully selected probability sample and a standardized questionnaire, and one can make descriptive assertions about any large population; (2) survey research makes it feasible to use large samples; (3) it offers more flexibility to the researcher in developing operational definitions based on actual observations; and (4) it generally uses standardized questionnaires, which from a measurement perspective offers strength to the data because the same question is being asked of all respondents, thus requiring credence to the same response from a large number of respondents.

Yet, despite the positives, the survey design also has its share of negatives: (1) a standardized questionnaire is limited with respect to whether the questions are appropriate because it is designed for all respondents and not for a select group; (2) it can seldom allow for the development of the feel for the context in which respondents are thinking or acting; (3) survey research is inflexible in that it typically requires that no change occur throughout the research, often requiring a preliminary study to be conducted; and (4) the survey tool is often subject to artificiality.

Even with these negatives, the survey design remains one of the most popular methods for conducting research. Furthermore, it is an extremely popular tool for collecting data. The issues involved in conducting survey research are discussed in more detail in later chapters.

Field Research

Chapter 5 discussed ethnography as a qualitative research method. In that discussion it was indicated that such research may often be quantitative in nature. When structured interviews are used that permit closed-ended questions to be answered, this becomes a personal interview. If field observations are made (e.g., observing how many vehicles ran a certain stop sign in a given time period or how many times members of the group being observed exhibited specific behaviors) allowing for numerical assignments, then this is quantified field research. When quantified, field research becomes an even more valuable tool for criminologic researchers.

FROM THE REAL WORLD

Standardized field sobriety tests (SFST) were developed during the 1980s to enable traffic officers to estimate accurately drug and alcohol impairment. These SFSTs consist of Walk-and-Turn, One–Leg Stand, and Horizontal Gaze Nystagmus. Traffic officers in all 50 states have been trained to administer these tests to individuals suspected of impaired driving and to score numerically their performance on them. It has been claimed that the SFST test battery is valid for detection of low blood alcohol concentrations and that no other measures of observation offer greater validity for blood alcohol concentrations of 0.08% and higher.

To study the legitimacy of these claims, Burns and Dioquino (1997) set up a field research in Pinellas County, Florida. Eight sheriff's deputies, with years of experience in traffic control, each having made hundreds of driving-under-the-influence arrests, and all having extensive training in driving-under-the-influence enforcement including certification in SFST, were selected to participate. A rigorous observation procedure was established in which both officers and observers assigned to each officer carefully recorded the SFST scoring and the actual outcome of measured blood alcohol concentrations. A total of 379 traffic stops were evaluated. In 313 cases SFSTs were administered. Drivers refused to take breath tests in 57 of these cases. In 256 cases, a breath test was administered to verify the accuracy of the SFST. The traffic officers were found to have an accuracy level of 97% in their arrest decisions based on the use of SFSTs.

Unobtrusive Research

Unobtrusive research does not require the researcher to be directly involved with the subjects of his or her study. There is no observation or interaction with the individuals or groups involved because the data have already been gathered by someone else or the data are available in a format that does not require such interaction. Examples of quantitative unobtrusive research include analysis of existing data, historical or archival research, and content analysis.

Analysis of Existing Data

Analysis of existing data is a very efficient way to conduct criminologic research. The data have already been gathered by a governmental organization, a research foundation, or an independent researcher. Rather than gathering new data, the researcher obtains and reanalyzes the existing data. This may include a reevaluation of the data from a prior study using a new method of analysis. Alternatively, it could be the use of government-generated data, such as Uniform Crime Reports or census data, in a new research analysis.

> **FROM THE REAL WORLD**
>
> To determine whether objective assessments of the risks and needs of court-involved youth lead to judicial processing that is less vulnerable to actual or perceived racial discrimination, Janku and Yan (2009) conducted a study for which the data came from the automated Missouri Justice Information System (JIS), which is maintained by the Office of State Courts Administrator. "Because of the wide differences across circuits in court location, court culture, and caseload size, we chose one midsized circuit in Missouri for this study to avoid potential noise created by those differences. The rationale for selecting this particular circuit is threefold: First, the caseload contains equal proportions of Caucasian and African American juveniles; second, the circuit contains a mix of metropolitan and rural areas; and third, the circuit has been on the JIS long enough to provide relatively reliable data" (p. 405).

Quantified Historical Research

Historical or archival research involves the review of prior research, documents, or social artifacts to gain insight about an event or era in history. Most of these types of research are qualitative. However, much of this research can be quantified, and empirical assessments are becoming more common on the part of historians and neo-Marxist criminologists.

Quantified Content Analysis

Like historical research, content analysis is the study of social artifacts to gain insight about an event or phenomenon. It differs in that the focus is on the coverage of the event by the particular medium being evaluated (books, magazines, television programs, news coverage, and so forth) rather than the event. Depending on how the research is conducted, this may be either qualitative or quantitative in nature. Quantitative content analysis emphasizes statistical rather than verbal analysis of various forms of communication. Instead of reading to understand the general emphasis or nature of the communication, the researcher either counts the number of occurrences in which a topic or issue is presented or numerically assesses the presentations based on a predetermined scale or ranking.

Evaluation Research

To this point, almost all the research designs discussed are used after an event, situation, or other unexplained phenomenon occurs and one wants to understand it better. These designs are quite useful to the academic researcher. What about the practitioner who wants to know how something will work or what might occur when something not previously done is attempted? In particular, what about the manager or administrator who is interested in making a change that could have sweeping policy implications? Often the best type of research design for this situation is evaluation research.

Evaluation research is a quantified comparative research design that assists in the development of new skills or approaches. It aids in the solving of problems with direct implications for the "real world." This type of research usually has a quasiexperimental perspective. For example, a police agency is debating whether to add a nonlethal weapon, a stun gun, to what is available to its officers. To see how these might work and the outcome of their use, an action design would use a select group of officers who are issued the stun gun for a set period of time. Each time it is used, a report explaining why and the results must be filed. At the end of the experimental period, the reports are analyzed and depending on the results, stun guns would be issued to all officers, certain officers, or not at all.

Evaluation research is important for the criminal justice practitioner–researcher because it assesses the merits of programs or policies being used (or under consideration for use) in the field. Whereas more basic research seeks to develop theoretical insights and much applied research seeks to determine if a theory can actually be applied in the field, evaluation research allows practitioners to determine the costs and effectiveness of the program or project that is being or has been implemented. For this reason, evaluation research that studies existing programs is frequently referred to as "program evaluation."

FROM THE REAL WORLD

To study the effects of environmental factors on a convenience store's vulnerability to robbery, Hunter (1988) conducted a statewide study of convenience stores. To conduct this study, a combination of quantitative research designs were used. From records provided by the Florida Department of Business Regulation 7740 convenience stores were found to be operating in the State of Florida during 1987. Two hundred stores were selected at random from this listing. Letters were sent from the Florida Attorney General to the heads of the law enforcement agencies within the areas where the stores were located. These letters contained a survey instrument seeking to obtain the number of times that each store had been robbed during the years 1984, 1985, and 1986. Hunter then visited each store to observe and rank them on a listing of environmental factors thought to be related to robbery vulnerability. Uniform Crime Report Data from the Florida Department of Law Enforcement were obtained regarding the crime rates, clearance rates, percent of sworn police personnel, and population numbers for each jurisdiction within which the sampled stores were located. By using a variety of statistical analyses, Hunter was able to identify several environmental factors as having an influence on a convenience store's vulnerability to robbery.

Combination Research

Each of the methods presented in this chapter is an excellent means of conducting quantitative research. However, there are even more strategies available. In criminologic research a combination of strategies often is used. A research design is, in actuality, a combination of qualitative and quantitative research in that a serious review of the prior research is expected to have been conducted, and its contributions to the development of the quantitative research. This is demonstrated in scholarly works in the problem statement and literature review.

Combinations of different quantitative methods are also commonly used. They may include the use of survey research and field observation, survey research and unobtrusive research, unobtrusive research and field observation, or a combination of all three. Overall, quantitative research offers a perspective that contrasts, yet can compare to, that which is discovered through qualitative research.

Summary

This chapter explores the nature of quantitative research by noting that it is research based on the assignment and assessment of numbers. Quantitative research is based on empiricism. It uses causality and probability to describe and explain relationships among variables. To construct a proper quantitative design, the level of measurement (nominal, ordinal, interval, and ratio) must be considered. Although there are many forms of quantitative research, the more popular include (1) survey research (personal interviews, mail questionnaires, and telephone surveys); (2) field research (quantified field observation); (3) quantified unobtrusive research (existing data, historical research, and content analysis); and (4) combination research, which uses a mixture of research methodologies.

METHODOLOGICAL QUERIES

1   You previously explained to the sheriff how the proposed study might work as a qualitative study. Now you must describe to him how it might work as quantitative research.

2   Since a quantitative study will produce various levels of data, identify and explain the four levels of measurement with applicable examples from the proposed research.

3   What type of survey would you use? Explain why.

4   How would you advise the sheriff with regard to the evaluative possibility of the proposed research?

SECTION II

Procedures

Research Designs

What You Should Know!

A conscious effort is required to conduct research and to use the most appropriate design. There are a multitude of methods available to conduct research and the method chosen by any researcher may be reflective of a style he or she is most comfortable with, believes will provide the best results, or is driven by the research question. This chapter examines the various designs from which researchers can choose. From this chapter the reader should be able to do the following:

1. Discuss the issues to consider in selecting a research design.
2. Describe how a historical research design is conducted.
3. Explain the merits of a descriptive research design.
4. Discuss developmental or time-series research. List the various types of time-series designs.
5. Compare and contrast longitudinal and cross-sectional research designs.
6. Compare and contrast trend studies, cohort studies, and panel studies.
7. Compare and contrast case studies with correlational and causal–comparative studies.
8. Discuss the strengths and weaknesses of experimental and quasi-experimental designs.

Research Designs

To complete any type of research successfully, it is important to establish a feasible plan or blueprint: the research design. This plan primarily responds to the common five Ws (who, what, where, when, and why) and H (how) of investigation. Because the justician and criminologist have a choice of research designs, it is important to be able to match the design properly to the desired outcomes. In selecting a research design there are a number of issues that should be considered. It is recommended that an outline be created to ensure that all relevant issues have been considered. Box 7-1 is an example of such an outline. Most of these issues are discussed in previous chapters. The remainder is covered in detail in later chapters.

This chapter discusses the more common forms of research designs applied to criminal justice and criminology. They include historical, descriptive, developmental, case, correlational, and causal–comparative. Brief mention is made of experimental and quasiexperimental action designs. The main focus is on what seems to be the most popular research design: survey research.

Historical

One of the most debated topics in criminal justice and criminology is the deterrent effect of capital punishment. A common hypothesis for pro–capital punishment supporters is that the death penalty serves as a better deterrent for homicides than life imprisonment. To study this hypothesis a historical, sometimes referred to as "records," research design is appropriate.

Historical designs allow the researcher systematically and objectively to reconstruct the past. This is accomplished through the collection, evaluation, verification, and synthesis of information, usually secondary data already existing in previously gathered records, to establish facts. The goal is to reach a defensible conclusion relating to the hypothesis. Using the hypothesis about the death penalty, a historical design could include a study of homicide rates in the United States between 1950 and 1997. Although just studying the numbers might provide an interesting conclusion, the true historical study requires

FROM THE REAL WORLD

To determine the impact on recidivism of matching youth with services at the individual level Vieira, Skilling, and Peterson-Badali (2009) conducted a study in which data were collected and compiled from a variety of secondary data sources. These sources included clinical charts; participants' court reports; each youth's scores on a risk and need measure (Youth Level of Service/Case Management Inventory); probation notes; parent and mental health service reports; and school records, probation files, and court records.

Box 7-1

Issues to consider in selecting a research design.

Purpose of research
  The purpose of the research project should be clearly indicative of what will be studied.
Prior research
  Review similar or relevant research. This promotes knowledge of the literature.
Theoretical orientation
  Describe the theoretical framework on which the research is based.
Concept definition
  List the various concepts that have been developed and clarify their meanings.
Research hypotheses
  Develop the various hypotheses that will be evaluated in the research.
Unit of analysis
  Describe the particular objects, individuals, or entities that are being studied as elements of the population.
Data collection techniques
  Determine how the data are to be collected. Who will collect it, who will be studied, and how will it be done?
Sampling procedures
  Sample type, sample size, and the specific procedures to be used.
Instruments used
  The nature of the measurement instrument or data collection device that is used.
Analytic techniques
  How the data will be processed and examined. What specific statistical procedures will be used.
Time frame
  The period of time covered by the study. This will include the time period examined by research questions and the amount of time spent in preparation, data collection, data analysis, and presentation.
Ethical issues
  Address any concerns as to the potential harm that might occur to participants. Also deal with any potential biases or conflicts of interest that could affect the study.

---

inclusion of possible intervening factors, which in this case could include US Supreme Court decisions, sentencing patterns, social episodes (e.g., a war), and population growth. A successful historical study considers all relevant information to provide proper conclusions.

The historical design is an economically efficient means for conducting research. Considering the vast array of records available related to criminal justice, there is no shortage of possible research topics. A shortcoming of this design is the difficulty of expanding beyond what is documented, therefore limiting the scope of the research. Researchers are limited to the information in the files and seldom have any means of following-up or getting clarification of the available information. This may render it unfit for the purpose of its use. In addition, there is the old computer maxim of "garbage in garbage out." The research is only as good as the data that is contained in the records. If there is inaccurate information or missing data in the original, the research suffers.

Descriptive

Assume one wants to know the composition of a particular jail population. A descriptive research design is very appropriate because it focuses on the description of facts and characteristics of a given population, issue, policy, or any given area of interest in a systematic and accurate manner. Like the historical design, a descriptive study can also rely on secondary or records data. It is also similar in that much historical research is descriptive in nature.

The information obtained in descriptive studies can provide insights not recognized in prior research. It can also lead to the study becoming inferential. An inferential study generalizes findings from a sample group and applies them to a larger population. However, descriptive design findings are not always accurate or reliable because of other elements not in the sample that are present in other parts of the population. Therefore, if one is going to make inferences from this type of study, they must be sure that the population is well represented within the sample studied.

Both historical and descriptive research can be cost-effective and logistically easier to conduct than other designs. However, they present researchers with limitations as to what variables can be examined and the extent of the information available. They may also be more time sensitive, which means that data may only be available for a certain time frame, and the information obtained may be limited in its usefulness. Thus, what if a researcher wants to control the information to be studied over a period of time that he or she controls? The answer is to use a longitudinal design.

Developmental or Time Series

Perhaps one is interested in following the activities of members from a rookie police class from graduation through their first five years of service. In particular the interest is in turnover rate, promotions, injuries, accommodations

FROM THE REAL WORLD

Adding to the literature on whether gang membership is uniquely related to victimization experiences for females compared to males, Gover, Jennings, and Tewksbury (2009) produced a descriptive in which they examined the relationship between gender, gang membership, and three types of victimization. They used data from the 1999 South Carolina Youth Risk Behavior Survey, an ongoing state and national survey conducted by state contracts for the Centers for Disease Control and Prevention, Division on Adolescent and School Health. Their results describe gang membership as significantly related to the risk of victimization regardless of gender.

FROM THE REAL WORLD

Violanti, et al. (2009) examined whether suicide ideation, planning suicide, and suicide attempts were more likely to occur among police officers because of their exposure to suicides. For their study they did a cross-sectional study involving 115 randomly selected police officers from a mid-sized urban police department of 930 officers. The sample was stratified by gender.

and complaints, and levels of job satisfaction. The best research design for this type of study is developmental, more commonly referred to as "time series" studies. This type of research design allows for the investigation of specifically identified patterns and events, growth, or change over a specific amount of time. Unfortunately, this type of research can be very costly and time-consuming so it is not readily used. There are several time-series designs that are available: cross-sectional studies, longitudinal studies, trend studies, cohort studies, and panel studies.

Cross-sectional Studies

The primary concept of the cross-sectional design is that it allows for a complete description of a single entity during a specific time frame. In this instance an entity might include an individual, agency, community, or in the case of the US Census an entire nation. These studies are best used as exploratory or descriptive research.

Longitudinal Studies

Whereas cross-sectional studies view events or phenomena at one time, longitudinal studies examine events over an extended period. For this reason, longitudinal studies are useful for explanation and exploration and description. Unfortunately, these types of studies are more difficult to conduct because they tend more often to be field research projects (Berg, 2008; Creswell, 2008).

Trend Studies

Trend studies examine changes in a general population over time. For example, one might compare results from several census studies to determine what demographic changes have occurred in that population. Surveys might indicate that opinions on the death penalty fluctuate over time depending on social conditions not related to crime and changes in the incidences of murder and the occasional occurrences of sensational murders.

Cohort Studies

Cohort studies are trend studies that focus on the changes that occur in specific subpopulations over time. Usually cohort studies use age groupings. For example, in a famous criminologic study, Wolfgang, Figlio, and Sellin (1972)

FROM THE REAL WORLD

Hunter and Wood (1994) were interested in the relationship between severity of sanction and unarmed assaults on police officers. They obtained assault data on officers for all 50 states. They compared these data with the sanctions applied for unarmed assaults on police officers within each respective state during 1991. They then compared the rates in states that had felony sanctions for weaponless assault on police officers to their neighboring states during the years 1977 through 1991. This resulted in the longitudinal analysis of four groups of states. Analysis of results in these groupings did not reveal support for the hypothesis that more sanctions would decrease the incidence of weaponless assaults on police officers.

examined delinquency among a cohort of juveniles. The findings from this study significantly impacted future research and practice in juvenile justice.

Panel Studies

Panel studies are similar to trend and cohort studies except that they study the same set of people each time. By using the same individuals, couples, groups, and so forth, researchers are able more precisely to examine the extent of changes and the events that influenced them. However, because of effects of deaths, movements from the area, refusal to continue as subjects, and other factors that cause the sample to lose members, these studies are logistically difficult to continue over an extended period of time.

Case Studies

The case (sometimes referred to as case and field) research design allows for the intensive study of a given issue, policy, or group in its social context at one point in time even though that period may span months or years (Adler & Clark, 2007). It includes close scrutiny of the background, current status, and relationships or interactions of the topic under study. Case studies often focus on a specific phenomenon, such as community policing.

Case studies may also be longitudinal in that they sometimes observe repeated cases over a certain length of time. These observations are closely linked with the observing of potential independent variables that may be associated with changes in the dependent variables. There are three basic features to this design (Dunn, 2009; Hagan, 2006; Maxfield & Babbie, 2009): (1) qualitative or quantitative descriptions of a variable over the extended period of time, (2) provides a context wherein the researcher can observe the changes in the variables, and (3) can be used for developing measurement instruments and the testing of their reliability over time.

Case research designs are not limited as to what can be studied. However, they can be costly and time prohibitive and may not provide an explanation

FROM THE REAL WORLD

Chappell's (2009) abstract states the following:

Community policing is the operating philosophy of the majority of American police departments in the new millennium. Though most departments claim to engage in community policing, research has shown that implementation of the strategy is uneven. One way to investigate the implementation of community policing is to study patrol officer attitudes toward community policing because research has shown that attitudes are related to behavior. The present study used qualitative data to explore the extent to which patrol officers have endorsed and implemented community policing in one medium-sized agency in Florida. Furthermore, the research sought to gain insight into the organizational barriers that prevented officers from adopting community policing in their daily work. Results indicated that although most officers agreed with the philosophy of community policing, significant barriers, such as lack of resources, prevented its full implementation in this agency. Implications of the findings and directions for future research are discussed. (p. 5)

for why the results turned out as they did. If one wants to know why something is or has occurred and possible correlating factors, then a correlational design may be more appropriate.

Correlational

A popular research design is one that allows researchers to investigate how one factor may affect or influence another factor, or how the one factor correlates with another—the correlational design. In particular, this type of design focuses on how variations of one variable correspond with variations of other variables. An example is a study of the level of education of police officers and promotion rates, arrest rates, or job satisfaction. The goal of this research design is to obtain correlational coefficients at a statistically significant difference.

Causal–Comparative

Why do men rape? Why do teenagers turn to gangs? Why do individuals become serial killers? To answer these types of questions, a causal–comparative (or ex post facto) design is useful. This design allows the researcher to examine relationships from a cause-and-effect perspective. This is done through the observation of an existing outcome or consequence and searching back through the data for plausible causal factors.

Each of the previously discussed designs can be found relatively frequently in the criminal justice and criminologic research. However, there are three other designs that could be used, but are much more difficult to use because of

FROM THE REAL WORLD

Interested in the relationship between drug addiction and criminal activity among incarcerated women who were in a prison drug rehabilitation program, Stevens' (1999) study could be viewed as a causal–comparative design. In this study, it was believed that drug addiction gave rise to criminality among females. Yet the results of the data did not support this belief. One implication of this finding was that drug addiction in itself is not necessarily a causal factor for producing crimes of violence, especially among females.

costs, logistics, and the fact that they cannot be easily applied to topics in criminal justice and criminology. They are the (1) true experimental, (2) quasiexperimental, and (3) action designs.

True or Classic Experimental

Although it is most often used in the natural sciences, occasionally social scientists may attempt research requiring a true experimental design. This type of design allows for the investigation of possible cause-and-effect relationships where one or more experimental units is exposed to one or more treatment conditions. The outcomes are then compared to the outcomes of one or more control groups that did not receive the treatment. This design includes four major components: (1) randomized assignment, (2) independent and dependent variables, (3) experimental and control groups, and (4) pretesting and posttesting.

The primary advantages of the experimental design are the isolation of the experimental variation and its impact over time; individual experiments can be limited in scope and require little time, money, and number of subjects; it is often possible to replicate the results. The major disadvantage is artificiality. Processes that occur in a controlled setting may not actually occur in the natural setting.

With respect to criminal justice and criminology, this type of research is often expensive and logistically difficult to perform. Probably one of the most difficult issues is obtaining consent when the research involves human test subjects. However, sometimes consent is easier to obtain if the experiment could prove useful to the subjects. For example, assume a new drug has been created that could suppress sexual desires. A group of convicted pedophiles are asked if they will participate in a study in which part of the group will receive the new drug, whereas the other half are given a placebo. After a certain number of weeks, both groups are tested for sexual response to certain stimuli. The results are compared.

Another major problem in conducting true experimental research is the difficulty in being able to maintain and control the environment where the experiment is conducted. The environment in which criminal justice and crim-

FROM THE REAL WORLD

To investigate the effectiveness of the Thinking for a Change, a widely used cognitive behavioral curriculum for offenders, Lowenkamp, Hubbard, Makarios, and Latessa (2009), evaluated the program through a group of offenders referred directly to the program from court. The results of their study found that the offenders who participated in the program had a significantly lower recidivism rate than similar offenders who had not been in the program.

inology research is conducted is often far from stable and filled with possible interfering or intervening variables. As a result of the control and consent issues, along with costs and other logistical problems, it is rare to see a justician or criminologist conduct true experimental research. This has not stopped some efforts to conduct this form of research. Some examples found in criminal justice include the Kansas City Preventive Patrol Experiment, the Minneapolis Domestic Violence study, and San Diego's one- versus two-person patrol units.

Reality dictates that few experimental research projects are possible in criminal justice and criminology. The limitations, however, can be addressed to some degree with a quasiexperimental design.

Quasiexperimental

Unlike the true experimental design where the researcher has almost complete control over relevant variables, the quasiexperimental design allows for the approximation of conditions similar to the true experiment. However, the setting does not allow for control or manipulation of the relevant variables.

Although easier to implement than the true experimental design, the quasiexperimental design has its difficulties, making it less appealing to most social scientists. The main difficulty lies in the interpretation of the results, that is, being able to separate the effects of a treatment from the effects caused by the initial inability to make comparisons between the average units in each treatment group. As with experimental designs, the quasiexperimental design is rare in theoretical criminologic research. It is more commonly used in evaluation research (discussed in Chapter 6) to evaluate new approaches in the criminal justice system and to solve problems with direct application to justice system operations.

Overall, there are a number of research designs from which to choose. The design chosen depends largely on what the researcher is seeking to discover, explain, or describe. Other considerations include economics, logistics, and time. Ultimately, the researcher must decide which design allows for the best results.

Summary

Selecting a topic and creating the research question are just the beginning of conducting research. One of the most important steps becomes the choosing of an appropriate research design. Although one of the most popular designs is survey research, which is also a means of collecting data in the other designs, there are a variety of other possible methods. These include the following:

Historical: reconstructing the past objectively and accurately, often in relation to the tenability of a hypothesis.

Descriptive: describing systematically a situation or area of interest factually and accurately.

Developmental: investigating patterns and sequences of growth or change as a function of time.

Case: studying intensively the background, current status, and environmental interactions of a given social unit, such as individual, group, institution, or community.

Correlational: investigating the extent to which variations in one factor correspond with variations in one or more other factors based on correlation coefficients.

Causal–comparative or "ex post facto": investigating possible cause-and-effect relationships by observing some existing consequence and searching back through the data for plausible causal factors.

True experimental: investigating possible cause-and-effect relationships by exposing one or more experimental groups to one or more treatment conditions and comparing the results to one or more control groups not receiving the treatment (random assignment being essential).

Quasiexperimentation: approximating the conditions of the true experiment, in a setting that does not allow the control or manipulation of all relevant variables.

The researcher must clearly understand what compromises exist in the internal and external validity of his or her design and proceed within these limitations. The chosen design should best meet the needs of the research goals.

METHODOLOGICAL QUERIES

1 By this point the sheriff is beginning to grasp the difficulty in conducting any type of study. He is aware that you think it would be better to conduct a survey, while he still favors observation. You decide to explain to him the issues to consider in selecting a research design.

2 Choose a design that you believe would best fit the proposed study and explain how it fits. Compare to at least one similar type of design.

Questionnaire Construction

What You Should Know!

The basis of research lies within the data. From where the data are obtained depends on the research design. One of the most popular means of gathering data is through a survey by way of a questionnaire. This chapter examines the use of questionnaires for collecting data. After studying this chapter the reader should be able to do the following:

1. Describe what is involved in listing the items one is interested in knowing about the group, concept, or phenomenon.
2. Explain how to establish validity and reliability.
3. Discuss why the wording in the questionnaire must be appropriate for the target audience.
4. Explain why who should answer the questions should be clearly identifiable.
5. Discuss why one should avoid asking any questions that are biased, leading, threatening, or double-barreled in nature.
6. Explain why before construction, a decision must be made whether to use open- or close-ended questions or a combination.
7. Discuss how it may be that the respondents may not have all the general information needed to complete the questionnaire.
8. Explain why a questionnaire should be pretested whenever possible before it is officially used.
9. Discuss why one should set up questions so that the responses are easily recognizable whether it is self-administered or an interview.

10. Explain why the questionnaire should be organized in a concise manner that keeps the interest of the respondent, encouraging him or her to complete the entire questionnaire.
11. Define scales and explain their purpose. Identify the different types of scales available for use.
12. Compare and contrast Thurstone Scales, Likert Scales, and Guttman Scales.

Questionnaire Development

When conducting survey research, a well-recognized research "rule of thumb" or perhaps the golden rule is to use a questionnaire that has previously been developed and tested. The primary reason for this is that it eliminates the worries of validity and reliability, two major concerns of questionnaire development. However, an instrument may not exist for a particular research question or, if it does exist, it may not meet the researcher's specific needs. Therefore, the researcher must resort to creating a research-specific questionnaire.

Rules for Questionnaire Construction

Many people prefer not to follow rules, or at least prefer to bend them to meet their satisfaction. The rules presented here do not have to be followed in a strict manner, but complete failure to follow them will lead to a failed questionnaire.

FROM THE REAL WORLD

While pursuing a doctorate degree in clinical psychology, Dantzker's dissertation examined whether there were any differences between two types of psychologists who performed pre-employment psychologic examinations of police recruits regarding what they used to conduct the evaluations and the reason for their choice. Because of the information sought, no existing questionnaire could accomplish this goal. Therefore, Dantzker (2010) created a study-specific questionnaire. The questionnaire used began from closed-ended questions answered through telephone interviews, developed into a self-administered questionnaire for a predissertation study (Dantzker & McCoy, 2006), and eventually was revised into the final question used for the dissertation.

In creating a new survey instrument there are several things to consider, including reliability and validity, and the level of measurement to use. Every textbook offers a different way to approach this task, but in the end, the basic elements are similar. To make understanding this task as easy as possible, a set of guidelines, offered in the manner of rules, is provided that can make questionnaire construction easier.

FROM THE REAL WORLD

In Dantzker and Waters (1999) study of students' perceptions of policing, the data for their sample was gathered in this manner and included the information shown in Figure 8-1. Listing the characteristics for the sample is the easier phase. Often researchers run into problems because they fail to list everything they know or want to know about the subject. This failure could cause a shortage of very important data. The key is to decide what is of interest and what is needed. The questions then take care of the rest, regardless of how they are asked, in statement form or as a question.

Ford and Williams (1999) used a short answer question approach in their survey instrument (Figure 8-2). Regardless of which approach is used, the key is to ask all the questions or create all the statements believed necessary to obtain the information desired. One can always eliminate data, but it is difficult to go back and get what was missed. In sum, Rule One suggests making a list of information desired before creating any questions or statements.

Rule One: Start With a List of All the Items One Is Interested in Knowing About the Group, Concept, or Phenomenon

How many times have you gone grocery shopping without a list? When that happens, it is common to end up with many things not wanted and missing things that are really needed. Questionnaire development should be approached with a grocery list mentality. List all the things you want to know, things you would like to know, and things that would be interesting to know about each item about which information is being sought.

Using Dantzker's dissertation topic, the primary interest was to examine whether there were any differences between psychologists as to the measures they use to conduct pre-employment psychologic screening. Also of interest was the reason for their choices. To collect the relevant information a questionnaire had to be developed that asked the respondents to provide certain identifying characteristics (the independent variables) along with their responses to questions about pre-employment screening (dependent variables). What was asked depended on the type of comparisons or analysis that was to be conducted. Because two groups were being compared, required information included information about the respondent and about the evaluation. Regardless of the nature of the sample, similar questions in various formats should be asked.

Rule Two: Be Prepared to Establish Validity and Reliability

After the golden rule, this may be the next most critical rule for questionnaire construction. These two concepts establish whether the data collected are acceptable to others who may be interested in using the findings, or perhaps even trying to replicate them.

Gender: Male ❑   Female ❑

Race/Ethnicity: White ❑   African-American ❑   Hispanic ❑
 Other _____

Age: _____

Year in College: Freshman ❑   Sophomore ❑   Junior ❑   Senior ❑   Other _____

Major: _____   **Minor:** _____

Last four digits of Social Security number (for analysis purposes only) _____

FOR CRIMINAL-JUSTICE STUDENTS ONLY

Other than Intro to Criminal Justice, have you taken any other courses with a police component? Yes ❑   No ❑   If Yes, how many?_____

Employment Goal: Law Enforcement ❑   Probation ❑   Courts ❑   Corrections ❑
 Law ❑   Other ❑

Are you related to a police officer? Yes ❑   No ❑   If Yes, what is the relationship?_____

Note the similarity in information, yet the differing format used in Ford and Williams' (1999) study of police and correctional officers' perception of a cultural-diversity course.

Personal Information

Sex: M ❑   F ❑

Race: American Indian ❑   Hispanic ❑   Asian ❑   White ❑   Black ❑
 Other _____

Age: 18–21 ❑   22–24 ❑   25–29 ❑   30–34 ❑   35–39 ❑   40–44 ❑   45–49 ❑   50+ ❑

Number of Years in Policing/Corrections: _____

Rank: Patrol _____   Supervisor _____

Ed: High School/GED ❑   2-year A.A. degree ❑   4-year B.S./B.A. degree ❑
Grad school or Degree ❑

Figure 8-1 Sample questionnaire.

Source: Reproduced from Dantzker, M. L. and Waters, J. E. Examining students' perceptions of policing—A pre- and post-comparison between students in criminal justice and non-criminal justice courses. In Dantzker, M. L., ed. *Readings for Research Methods in Criminology and Criminal Justice.* Butterworth-Heinemann. Copyright Elsevier 1999.

1. When did you take the human diversity course? Year_____
2. How would you rate your understanding of other cultures before taking the course?
3. How would you rate your understanding of other cultures after taking the course?
4. Has there been a difference in your on-the-job behaviors (beliefs, the way you think about others, or your actions) since you took the course? Can you give an example? [Use the reverse side if you need more room.]
5. Has there been a difference in your off-duty behavior or your personal life since you took the course? Can you give an example? [Use the reverse side if you need more room.]
6. What do you believe is the single most important thing you learned from the course? [Use the reverse side if you need more room.]
7. What, if anything, was the worst thing about the course? [Use the reverse side if you need more room.]
8. Would you recommend/have you recommended the course to others?

Figure 8-2 Short Answer Question Approach

Validity refers to whether the questionnaire is in fact measuring what it claims to measure. It is imperative that the questions or statements be a true measure of the topic under study. This requires that the researcher be able to determine and establish the questionnaire's validity, which can be accomplished through one or more of the following ways: face, content, construct, and criterion validity.

Face

The simplest means of establishing validity is face validity. This requires the researcher to accept that the questionnaire is measuring what is being attempted to be measured because the researcher believes it is. This makes it a very judgmental process, lacking empirical support, and requiring the researcher to demonstrate why he or she believes it measures what is expected. The statements are often developed based on the researcher's knowledge, background, and observational experiences. Although this is an acceptable means of validation, it is the least acceptable because it is empirically weak.

Content

The second form of validation, content validity, also suffers from being judgmental and usually nonempirical. Unlike face validity, where the belief in validity focuses on the questionnaire as a whole, content validity emphasizes each individual item's ability to measure the concept in question. Instead of simply supporting the complete questionnaire's ability to measure what is expected, the researcher must be able to explain why each item measures what is expected. Again, the responsibility falls to the researcher to support why it is believed that each item measures what is expected.

Construct

Although one is seeking to measure a particular phenomenon, there may be related concepts that are equally important to understanding the phenomenon in question. Construct validity (sometimes referred to as "concept validity") seeks to demonstrate that the questions actually measure what they have been designated to measure in relation to other variables. The interest is in establishing the fit between the theoretical and operational aspects of the item. With respect to the job satisfaction study, it was expected that individuals with a college degree would indicate a higher level of satisfaction with some items than individuals without a college degree. If the responses to those particular questions support this, then one has established construct validity. However, if the responses from both groups indicate equal satisfaction, then the validity may be challenged. Construct validity can be reinforced through empirical measures.

Criterion

Criterion, also called "pragmatic" or "empirical," validation is concerned with the relationship between the questionnaire and its results. The assumption is that if the questionnaire is valid, a certain empirical relationship should exist between the data collected and other recognizable properties of the phenomenon. Most often the evidence to support this is garnered from correlational measures consistent with the level of measure. The key requirement, however, is that a reliable and valid measure must already exist to make the comparison. To apply criterion validity to the previously mentioned job satisfaction measure, one group of officers is given the job satisfaction measure, which is then compared to the results of a reliable and validated job stress measure (because it has been established that job stress is linked to job satisfaction). With the two sets of scores, a correlation coefficient can be computed providing what is called the "validity coefficient." The more common form of criterion validity is predictive validity, which rests on the questionnaire's ability accurately to predict future conditions or responses.

There may be debates over the best type of validity. Failing to establish any type of validity devalues the data. The more ways one can establish validity, the better. Still, validity alone is not enough. It must be accompanied by reliability.

Everyone who lives in a climate where there are several days of extremely cold weather wants a car battery that starts the vehicle day in and day out. Perhaps the battery was purchased because of its reliability to do just that. When it does not start as expected, it is no longer considered reliable. A questionnaire has the same expectation: that it is reliable each time it is used. If the questionnaire is consistent over time and yields similar results each time it is used, it is reliable. To establish reliability further, one must demonstrate stability and consistency.

FROM THE REAL WORLD

As previously noted, Dantzker (2010) created a questionnaire specifically for his research. For this study's questionnaire face validity was addressed through the researcher's knowledge of the subject, comparison to previous research efforts, and support from neutral observers with questionnaire construction experience and knowledge of the subject. The questionnaire also met the criteria for content validity. The content of the questionnaire was primarily the identified protocols and reasons for choosing them, both of which have been established by relevant literature. The questionnaire was reviewed by three licensed psychologists who all indicated they found that the questionnaire does reflect the intended objectives. Finally, for criterion and construct validity the questionnaire had to cover the full range of possibilities within the concept, and required a pre-existing questionnaire with which to compare or a concept that cannot be directly observed or isolated. Because the concepts measured were based on actual use and established reasons for its use, the questionnaire did measure the primary concept.

Stability occurs when, under similar conditions, a respondent provides the same answers to the same questions on a second testing. Consistency is determined when the set of questions is strongly related and is measuring the same concept. There are three standard ways to test reliability: (1) test–retest (pretesting); (2) split-half technique; and (3) using multiple forms.

Pretesting is the most fundamental method, yet perhaps the most inconvenient in terms of time and money. The test–retest method requires distributing the questionnaire to the same population twice. If the results are the same, then reliability is accepted. Another method is to distribute the questionnaire to similar samples and look for consistent results between the samples.

A popular and widely used method is the split-half technique. Here the questionnaire is divided into sections or halves. Both sections are given to the same group or among similar groups. A similarity of scores between both halves supports stability.

Using several variations or formats of the same questionnaire, the multiple forms method can support stability. As with the previous methods, if scores on each format are similar, one can assume stability.

All these methods are acceptable; however, when possible one should use the test–retest method. In addition, many statistical packages offer methods for statistical comparison by item-to-item and item-to-scale analyses, and the use of Cronbach alpha, a commonly used reliability coefficient.

Overall, there are a number of ways to establish validity and reliability. Therefore, there is no reason to fail to meet Rule Two. The goal is to be able to establish both validity and reliability.

Rule Three: The Wording in the Questionnaire Must Be Appropriate for the Target Audience

Everyone is required to complete questionnaires of one type or another at some point in their lives. Sometimes the questions are quite clear; other times they may befuddle the respondent. When developing a questionnaire, the first guideline is to be sure to use language geared toward the target population. One should not use words or phrases with which the respondent is not familiar; it can cause confusion and misunderstanding and may lead to tainted data. For example, on a questionnaire for college students and criminal behavior, a question about drug use should use the slang or more common usage rather than scientific: Have you ever smoked cannabis? (wrong); Have you ever smoked marijuana? (right). Rule Three suggests that the questions or statements be written in a manner that the target audience can understand.

Rule Four: Be Sure That It Is Clearly Identifiable Who Should Answer the Questions

How many times has a questionnaire in the mail been addressed to Dear Occupant, and on opening it one discovers that it is not clear who should be completing this questionnaire, or it is not be clear who should fill out which parts. For his questionnaire, Dantzker (2010) included this question and response instructions to clarify who should continue completing the questionnaire:

> *Do you personally provide psychological screening services to any type of police/law enforcement agency?*
> *Yes* ❑   *IF YES, please respond to the remainder of survey.*
> *No* ❑   *IF NO, you may stop here. Thank you.*

All it takes is a simple statement advising who should complete which questions. Thus, for Rule Four the idea is to be sure to clarify who fills out or responds to which questions.

Rule Five: Avoid Asking Any Questions That Are Biased, Leading, Threatening, or Double-Barreled in Nature

"Does it not feel great to get high?" "How often do you get high and do you enjoy it?" "Do you cheat on exams and if you do, do you know you are only cheating yourself?" Questions like these are considered biased, double-barreled, or leading. The basic premise is not to create questions where there is confusion for the respondent on how to answer, nor is there a push toward a particular answer.

Questions or statements that confuse respondents can cause ambiguous responses. Questions that seek to guide the respondent can create blatantly false responses. In addition, the structure of the questionnaire may be such that preceding questions influence the responses to later questions. If any of this occurs, the findings are not valid.

Ultimately, the wording of questions and questionnaire format can influence responses. Be aware of that when responding to surveys conducted by ideologic groups or organizations that hold particular positions on controversial issues (e.g., the opposing views of Handgun Control and the National Rifle Association). Potential biases may also be observed among the questionnaires of scholars and students who are trying to support a favored hypothesis. If those administering a survey have an interest in or might benefit from the outcome of the survey, they may word the questions or structure the format so as to enhance the likelihood of desired responses. Be skeptical of survey findings by such individuals or groups.

Rule Six: Before Construction, a Decision Must Be Made Whether to Use Open- or Closed-Ended Questions, or a Combination

Because the goal of the questionnaire is to ascertain specific information related to the topic and readily to analyze this information, deciding on what type of questions should be asked is important. Open-ended questions can make data analyses somewhat more difficult, but can provide more in-depth responses. On the other hand, well-constructed closed-ended questions can provide sufficient data that are more readily analyzable. The more popular method is to combine the types of questions to collect the most pertinent information about the topic.

The questionnaire in Figure 8-3 was created as a telephone survey to investigate community satisfaction with a police department. Although this questionnaire had been created as part of a proposal to evaluate a police agency, it was never tested or published elsewhere. Observe how the questionnaire is composed of both closed- and open-ended questions. Notice how the open-ended questions are worded so that responses could be more readily coded for statistical analysis.

Rule Seven: Consider That the Respondents May Not Have All the General Information Needed to Complete the Questionnaire

Under any circumstances, making assumptions could be problematic. This is especially true in questionnaire development. It is a common error to believe that would-be respondents have all the information needed to respond to the questionnaire. For example, several questions about drug and alcohol programs on campus might be asked in the Student Criminality study simply because it is assumed that all students are familiar with these programs. This assumption might result in few responses, causing another incorrect assumption about the results. To avoid this dilemma, always provide an "escape" response such as "unknown" or "no prior knowledge." Refer back to Figure 8-3 and the use of the "unable to respond" choice.

POLICE COMMUNITY TELEPHONE SURVEY

1. Have you had any official contact (by telephone or in person) with any member of the XXXXX Police Department during the past 18 months?   ❑ Yes   ❑ No

IF YES, what was the reason/circumstance?

❑ Traffic stop   ❑ Traffic accident   ❑ Victim of a crime

❑ Arrested   ❑ Witness to accident   ❑ Witness to a crime

❑ Informational   ❑ Telephone contact   ❑ Other

How would you rate the performance of the person you had contact with?

	Wholly Unsatisfactory	Less than Satisfactory	Adequate	Completely Satisfactory	Unable to Respond
	1	2	3	4	0
Providing Assistance	1	2	3	4	0
Knowledge	1	2	3	4	0
Courtesy	1	2	3	4	0
Sensitivity	1	2	3	4	0
Friendliness	1	2	3	4	0
Handling of Situation	1	2	3	4	0
Overall Conduct	1	2	3	4	0

IF NO, what is your opinion, perception, or attitude of the XXXXX Police:

	Wholly Unsatisfactory	Less than Satisfactory	Adequate	Completely Satisfactory	Unable to Respond
	1	2	3	4	0
Providing Assistance	1	2	3	4	0
Courtesy	1	2	3	4	0
Sensitivity	1	2	3	4	0
Friendliness	1	2	3	4	0
Handling of Situation(s)	1	2	3	4	0
Competence	1	2	3	4	0
Attitude	1	2	3	4	0
Behavior	1	2	3	4	0
Traffic Enforcement	1	2	3	4	0
Crime Prevention	1	2	3	4	0
Enforcing Laws	1	2	3	4	0
Solving Crimes	1	2	3	4	0
Dealing with Citizens	1	2	3	4	0
Dealing w/Arrested Persons	1	2	3	4	0
Overall Performance	1	2	3	4	0

2. In your opinion, what is the main problem needing police attention in your neighborhood?

3. In your opinion, what is the main problem needing police attention in the city of XXXXX?

Figure 8-3 Questionnaire: Open- and closed-ended questions.

POLICE COMMUNITY TELEPHONE SURVEY

4. In your opinion, what is the BEST thing about the XXXXX Police?

5. In your opinion, what is the WORST thing about the XXXXX Police?

6. In your opinion, what is one change you would make that could improve the XXXXX Police?

For Analysis Purposes Only:
Gender   ❑ Male   ❑ Female

Race/Ethnicity ❑ Caucasian   ❑ African-American   ❑ Hispanic   ❑ Asian   ❑ Native American
❑ Other

Age   ❑ 18–29   ❑ 30–49   ❑ 50–65   ❑ Over 65

Employment ❑ Professional (i.e., doctor, lawyer, etc.)   ❑ Education   ❑ Retail Business
❑ Food Service   ❑ Clerical   ❑ Laborer   ❑ Other

Education ❑ High School   ❑ Some College   ❑ Two-year Degree   ❑ Four-year Degree
❑ Graduate Degree

In what area of the city do you live?
❑ North   ❑ South   ❑ East   ❑ West   ❑ Central

THANK YOU FOR YOUR ASSISTANCE.

Figure 8-3 Questionnaire: Open- and closed-ended questions. (continued)

Rule Eight: Whenever Possible, Pretest the Questionnaire Before It Is Officially Used

A reinforcement to Rule Two (stability), and one of the most difficult of the rules to follow, this rule is undoubtedly very legitimate and should be used whenever possible. With the Student Criminality questionnaire example, after completing the first draft of the questionnaire one might have members in one's classes complete the questionnaire. Pretesting can find errors in construction, language, or other errors that may cause the data to be useless if not corrected. Keep in mind, this does not necessarily require anything more than a few individuals from the target population willing to complete the questionnaire and provide feedback. Although this process might take a little more time and effort, it is well worth it in the long term.

FROM THE REAL WORLD

Returning to Dantzker's (2010) questionnaires, specific instructions were provided and the way and means for answering were quite clear (Figure 8-4).

The purpose of this survey is to garner a clearer understanding of how psychologists who screen potential police recruits label themselves and how they may differ by label as to what it is they use to conduct the screening and why they use what they do. No personal identifiers are required. It should take less than 15 minutes to complete. Your cooperation is appreciated. Thank you!

Male ❏   Female ❏   State in which you practice _____

Years in practice as a Clinical Psychologist: < 1 ❏   1–5 ❏   6–10 ❏   > 10 ❏

Association Membership: APA ❏   IACP ❏   Both ❏

1. Does your state mandate psychological assessments for individuals being considered for employment with a police agency?

   Yes ❏   No ❏   I don't know ❏

   If "Yes": Does your state mandate the use of a particular assessment tool or tools?

   Yes ❏   No ❏   I don't know ❏

   If "Yes," please list the name(s) of the tool(s)_____

2. Do you personally provide psychological screening services to any type of police/law enforcement agency?

   Yes ❏   IF YES, please respond to the remainder of survey.
   No ❏   IF NO, you may stop here. Thank you.

3. How do you identify or label yourself with respect to providing police screening services?

   ❏ Police Psychologist (full time in-house psychologist)
   ❏ Psychologist Consultant (full-time consultant to law enforcement)
   ❏ Clinical Psychologist (occasional service provider to law enforcement)
   ❏ Other (please describe) _____

4. Years conducting police pre-employment screenings

   < 1 ❏   1–5 ❏   6–10 ❏   > 10 ❏

5. To what types of agency do you provide services? (Please check all that apply)

   ❏ Municipal (city, village, or similar agency)
   ❏ County or parish (sheriff, or similar agency)
   ❏ State
   ❏ Other (Please identify type) _____

Figure 8-4 Sample questionnaire with specific instructions.

Rule Nine: Set Up Questions So That the Responses Are Easily Recognizable Whether It Is Self-Administered or an Interview

The fastest way to jeopardize research is through a questionnaire in which the respondents are not clear on how to respond. Be sure to provide adequate, clear instructions and establish recognizable means for responding. Also, try not to make the format too busy (hard to read because too many questions are squeezed on a page). One should avoid using small print.

Rule Ten: The Questionnaire Should Be Organized in a Concise Manner That Keeps the Interest of the Respondent, Encouraging Him or Her to Complete the Entire Questionnaire

How often have you started a novel only to not make it past the first few chapters because it was boring? If you had continued to read, it might have become more pleasurable, but you had lost interest. This same concept is applicable to questionnaire development. If the beginning questions are not interesting and do not hold the respondents' attention, chances are that they may not complete the rest of the questions, which results in missing data. Therefore, it is beneficial to have questions that may pique the respondent's interest in the beginning and at the end. In addition, if respondents see several pages of questions, they are less likely to begin the survey. Although it is tempting to try to cover everything, a clear and concise survey that consists of a few easy-to-read questions receives more responses. Although specialized questionnaires that target a specific population (that has a vested interest in the subject matter) may be longer, it is recommend that most surveys be kept to two pages of questions using normal-size type.

Rules and guidelines are fallible, but the 10 rules offered for questionnaire construction, if followed, improve the chances of obtaining good, analyzable data. Still, keeping all the rules in mind might not prevent the creation of a poor questionnaire. Consequently, these rules are not the only thing that must be known to create a usable questionnaire. A key aspect in questionnaire construction is measurement.

Scales

A common element of survey research is the construction of scales. A scale can either be a measurement device for responding to a question or statement or a compilation of statements or questions used to represent the concept studied. For example, in the questionnaire in Figure 8-3, responses to several statements are a scale that ranges from one to four. Each statement is a separate variable of the respondent's perception. Putting the statements from each perceptual question together (attaining a numerical result for responses to all statements) gives the researcher what might be referred to as the "Community Member's Perceptions of Their Local Police Agency Scale." Scales as compilations are particularly important and a relevant part of research for three primary reasons (Creswell, 2008; Dunn, 2009; Frankfort-Nachmias & Leon-Guerrero, 2008): (1) to allow the collapsing of several variables into a single variable producing a representative value, thus reducing the complexity of the data; (2) to offer measures that are quantifiable and more open to precision and statistical manipulation; and (3) to increase the measurement's reliability.

To accomplish these things, a scale must fit the Principle of Unidimensionality (Frankfort-Nachmias & Leon-Guerrero, 2008). This principle suggests that the items making up the scale need to represent one dimension befitting a continuum that is supposed to be reflective of only one concept. For example, if one is measuring job satisfaction, the scale should not be capable of also measuring job stress. The representativeness of any scale relies greatly on the level of measurement used.

Scaling Procedures

Ultimately, to conduct research one looks to complete a measurement. Yet, what is the actual purpose of this measurement? Measurement is used in research as a means of connecting phenomena with numbers for analytical purposes. Scaling is identified as a means of assisting in making the necessary and proper connections. Despite the existence of numerous scales, there are times when the researcher must create his or her own. The key is to understand that one is trying to explain a phenomenon and that the scale must meet this criterion. There are two primary types of scaling procedures from which to choose.

Arbitrary Scales

An arbitrary scale is designed to measure what the researcher believes it is measuring and is based on face validity and professional judgment. Although this allows for the creation of many different scales, it is easily criticized for its lack of substantive support. Still, this type of scale does provide a viable starting point for exploratory research, even though it is the less recommended method of scaling.

Attitudinal Scales

More commonly found in criminal justice and criminologic research are the attitudinal scales. There are three primary types available: (1) Thurstone, (2) Likert, and (3) Guttman.

Thurstone Scales

The construction of a Thurstone scale relies on the use of others (sometimes referred to as "judges") to indicate what items they think best fit the concept. There are two methods for completing this task. The first method is paired comparisons. Here, the judges are provided several pairs of questions or statements and asked to choose which most favorably fit the concept under study. The questions or statements picked most often by the judges become part of or comprise the complete questionnaire. For example, the judges are asked to which of the following questions might best fit the concept of job satisfaction: "I enjoy going to work every day." "Sometimes I am very tired when I get home from work."

The second method and one more often used is equal-appearing intervals. For this method, the researcher submits a list of questions or statements to the judges who are then asked to give each a number depending on how large a scale is desired, indicating the strength of the question or statement to the concept. The researcher then keeps those items where judges were in the strongest agreement and eliminates those with the weakest indicator scores. For example, the researcher decides to design a questionnaire to examine criminality among college students and calls it the "College Students' Criminality Questionnaire." A decision is made to have the questions form a 15-point item criminality scale. Fifty questions are submitted to judges asking them to score each question from 1 (strongest indicator) to 15 (weakest indicator). The top 15 questions become the scale.

Thurstone scaling is not very popular because of the time it takes for the judges to complete their tasks. Furthermore, because the judges have to be experts in the area of study, finding an adequate pool of qualified judges could also be problematic for the researcher.

Likert Scales

Probably the most commonly used method in attitudinal research is the Likert scale. This method generally makes use of a bipolar, five-point response range (i.e., strongly agree to strongly disagree). Questions where all respondents provide similar responses are usually eliminated. The remaining questions are used to comprise the scale. Figure 8-5 provides an example of Likert-type questions.

Guttman Scales

The Guttman scale requires that an attitudinal scale measure only one dimension or concept. The questions or statements must be progressive so that if the respondent answers positively to a question, he or she must respond the same to the following question.

There are various other types of scaling procedures. However, because so few are used in criminal justice and criminologic research they are not discussed here. Furthermore, advanced statistical techniques, such as factor analysis and Chronbach alpha, are much faster and simpler to use to determine a scale's composition. For example, the Student's Perceptions of Police scale (Dantzker & Waters, 1999) was originally 20 items before a factor analysis and Chronbach alpha eliminated six items, bringing it to its final 14-point scale. The question at this point is, why use scales at all?

There are three primary reasons or advantages to using scales. First, a scale allows for a clearer and more precise measure of the concept than individual items. Second, scales can be replicated and used as longitudinal measures.

Using the following scale, please circle the number that best describes your response to each statement.

	Strongly Agree	Agree	Doesn't Really Matter	Disagree	Strongly Disagree
	5	4	3	2	1
1. Premarital sex is acceptable for males.	5	4	3	2	1
2. Premarital sex is acceptable for females.	5	4	3	2	1
3. Oral sex before marriage is acceptable.	5	4	3	2	1
4. Oral sex is deviant behavior and should never be practiced.	5	4	3	2	1
5. Females should be virgins at the time of their marriage.	5	4	3	2	1
6. Males should be virgins at the time of their marriage.	5	4	3	2	1
7. A male should have some type of sexual experiences prior to being married.	5	4	3	2	1
8. A female should have some type of sexual experiences prior to being married.	5	4	3	2	1
9. People should first live together prior to getting married.	5	4	3	2	1

Figure 8-5 Example of Likert scale statements.

Source: Reproduced from Eisenman, R., & Dantzker, M. L. (2006). Gender and ethnic differences in sexual attitude at a Hispanic-serving university. *Journal of General Psychology*, 133 (2): 153–162. Reprinted by permission of the publisher (Taylor & Francis Ltd, http://www.tandf.co.uk/journals).

Finally, scales require more thought. The disadvantages to scales are twofold: there is concern as to whether true attitudes can be measured on a scale, and there is the question of validity and reliability. Despite the shortcomings, overall, scales can be quite useful in measuring data and should be used where and when it is appropriate and necessary.

Summary

To conduct research, data must be collected. Generally, this means that some type of tool (a questionnaire) must be available to assist in collecting the data. The golden rule is to try to use a questionnaire that has previously been tested. However, when that is not possible and a questionnaire must be constructed, following the suggested rules helps create an acceptable tool. These rules are as follows:

1. Start with a list of all the items one is interested in knowing about the group, concept, or phenomenon.

2. Be prepared to establish validity and reliability.

3. The wording in the questionnaire must be appropriate for the targeted audience.

4. Be sure that it is clearly identifiable as to who should answer the questions.

5. Avoid asking any questions that are biased, leading, threatening, or double-barreled in nature.

6. Before construction, a decision must be made whether to use open- or closed-ended questions, or a combination.

7. Consider that the respondents may not have the general information needed to complete the questionnaire.

8. Whenever possible pretest the questionnaire before it is officially used.

9. Set up questions so that the responses are easily recognizable whether it is self-administered or interview.

10. The questionnaire should be organized in a manner that keeps the interest of the respondent, encouraging him or her to complete the entire questionnaire.

In addition to the rules, questionnaire development requires familiarity with such issues as reliability, validity, measurement level, and scales. Scales can be either arbitrary or attitudinal in nature. Three primary attitudinal scales are Thurstone, Likert, and Guttman. The most popular scale is Likert, which makes use of a bipolar set number of points.

METHODOLOGICAL QUERIES

1 You have convinced the sheriff that the study of job satisfaction should be conducted through a survey. Now he wants to know what you would include in the questionnaire. Describe what you are interested in knowing about the group, concept, or phenomenon.

2 Since you believe an appropriate questionnaire doesn't exist for measuring job satisfaction among correctional officers, you explain to the sheriff how you will have to create your own. He wants to know what that will entail. Explain:

   a the need to establish validity and reliability;

   b the need for the wording in the questionnaire being appropriate for the target audience;

   c why it should be clearly identifiable who should answer the questions;

   d how you'll avoid asking any questions that are biased, leading, threatening, or double-barreled in nature;

   e why prior to construction, a decision must be made whether to use open- or close-ended questions, or a combination; and

   f why a questionnaire should be pretested whenever possible before it is officially used.

3 What type of scale will you use? Explain why.

Sampling

What You Should Know!

Gathering the data as discussed in Chapter Eight can be done in a variety of ways. Using a questionnaire is very popular. However, from whom the data are collected becomes important. Often data cannot be collected by every member of a target population. Therefore, sampling is necessary. After studying this chapter the reader should be able to do the following:

1. Discuss the purpose of sampling in criminologic research.
2. Define what is meant by population, sampling frame, and sample. Describe their relationships. Give examples of each.
3. Explain how probability theory enables the researcher to obtain representative samples.
4. Identify and describe the various types of probability samples.
5. Compare and contrast probability sampling with nonprobability sampling.
6. Identify and discuss the various types of nonprobability samples.
7. Explain the importance of sample size. Include confidence intervals, confidence levels, and sampling error in the discussion.
8. Determine how many observations are necessary to obtain a sample with an error tolerance of ±3 at the 95% confidence level. Explain how many more observations you would add to this number and why you would do so.

Sampling

Conducting research requires the gathering of information about a specific concept, phenomenon, event, or group. Although there are some types of research that require gathering information about every element associated with the topic, a natural science necessity, in the social sciences, this is neither feasible nor necessary. In conducting criminal justice and criminologic research the primary focus is usually on some population.

Recall from Chapter 3 that a population is the complete group or class from which information is to be gathered. For example, police officers, probation officers, and correctional officers are each a population. Again, although it would be great if every member of a population could provide the information sought, it is just not practical. Therefore, sampling the population is a necessity. Before one can begin to sample, the first step is to identify the group from which the sample will come. This is called the "sampling frame," which if the sample is to be representative must include as close to the complete population as possible (Bachman & Schutt, 2008; Creswell, 2008).

Having identified the sampling frame, the next decision is to choose the type of sample to be used. Again, recall from Chapter 3 that a sample is a group chosen from a target population to provide information sought. That sample is either of the probability or nonprobability nature. It is appropriate at this point to provide an overview of probability theory.

Probability Theory

A friend of ours frequently buys lottery tickets. On a number of occasions individuals (usually whom he does not know) have taken it on themselves to inform him that, "You are wasting your money. The odds of you being hit by lightning are higher than of you winning the lottery." To which he honestly replies, "Thank you for your concern. I have been hit by lightning. This is more fun."

FROM THE REAL WORLD

Let us continue with Dantzker's dissertation examining whether there were any differences between two types of psychologists who performed preemployment psychologic examinations of police recruits regarding what they used to conduct the evaluations and the reason for their choice. Because there is no one source identifying all licensed psychologists in the United States, the sampling frame for this study included all clinical psychologists with memberships in the American Psychological Association (APA) and the Psychology Section of the International Association of Chiefs of Police.

The individuals who are warning our friend are statistically correct: his chance of winning the lottery is extremely small and actually is less likely than being hit by lightning. However, because he does not really care if he wins (he amuses himself by checking them several days later, asserting that he is potentially a winner until he discovers otherwise) and because his investment only averages about $2 per week, statistical probabilities do not mean much to him.

Unfortunately, many people who wager far more than they can afford also disregard statistical probability. These individuals tend to believe either that "it is time for their luck to change" or that "God (or fate, depending on their religious orientation) will intervene in their lives." Although we do not question the benevolence of a Supreme Being, we do believe that if God or even luck preordained such an event, the purchase of one ticket would be adequate. If not preordained, the sincere gambler might want seriously to consider the statistical probability of success.

Probability theory is based on the concept that over time there is a statistical order in which things occur. If one flips an unaltered coin 10 times, it is possible that it will land on heads five times and tails five times. However, it might land on heads eight times and tails only two. This is because each time the coin is flipped it has an equal chance of being heads or tails. What happened previously has no influence on what happens in the future. One cannot accurately predict what will happen on the next coin toss. Yet, one can accurately assume that over a lengthy period of time the number of heads and the number of tails will be about the same. This is the basis of statistical probability. Anything can happen, but over the long run there is a statistical order.

The knowledge that over time things tend to adhere to a statistical order allows one to choose samples that are representative of a population in general. Although one cannot say in advance that a sample is representative, one can follow a procedure that should lead to a representative sample being selected. Because every number (representing people, items, or events) has the same chance of being chosen, then most of the time the sample drawn is representative. On occasion it is not.

Probability Sampling

The general goal when choosing a sample is to obtain one that is representative of the target population. By being representative, the results can be said to be applicable to the whole population and are similar no matter how many different samples are surveyed. Representation requires that every member in the population or the sampling frame have an equal chance of being selected for the sample. This is a probability sample. Four types of probability samples exist: (1) simple random, (2) stratified random, (3) systematic, and (4) cluster.

Simple Random Samples

A random sample is one in which all members of a given population have the same chance of being selected. Furthermore, the selection of each member must be independent from the selection of any other member. To assist in selecting a random sample, a device known as a "table of random numbers" is often used, which can be found in almost every statistics book. One selects a numeral at random and then uses the subsequent numerals provided within the table until the appropriate number of population members needed has been selected. Today, using the computer to generate a sample randomly is becoming more popular. In either case, the researcher must have a complete list of every member of the sampling frame, which is one of the disadvantages of random sampling. Yet, even with this obstacle, random sampling is very popular, partly because it has become easier for researchers to obtain statistically acceptable, random samples.

Stratified Random Samples

Chapter 3 defines a stratified sample as one that has been chosen from a population that has been divided into subgroups called "strata." These strata are selected based on specified characteristics that the researcher wishes to ensure for inclusion in the study (Adler & Clark, 2007; Dunn, 2009; Frankfort-Nachmias & Leon-Guerrero, 2008). This type of sample requires the researcher to have knowledge of the sampling frame's demographic characteristics. These characteristics (selected variables) are then used to create the strata from which the sample is chosen. Depending on the interests or needs of the researcher, a proportionate selection is made from each stratum (in a random manner) or oversampling (disproportionate) may be necessary.

Systematic Samples

There is some debate over this type of sampling. It has been discussed as both probability and nonprobability sampling. It is offered here as a probability sample because it includes random selection and initially allows inclusion of

FROM THE REAL WORLD

Examining the shadow of sexual assault hypothesis among college and university students across temporal situations and victim-offender relationships, Hilinski (2009) used a simple random sample. The sample was of 375 undergraduate and graduate students enrolled at a medium-sized public university during both the fall 2006 and spring 2007 semesters. She obtained her sample by sending an e-mail invitation to a random selection of 3500 (approximately 25% of the university's population) graduate and undergraduate students enrolled during the spring 2007 semester.

> **FROM THE REAL WORLD**
>
> Dantzker used a combination of sampling techniques for his study. He began with a stratified, systematic random sample drawn from the membership of APA Division 12 (Clinical Psychologists). After removing students, non–United States members, and those who retired, a population of 3609 members was left. A random sample of 1000 was selected through stratification by state. Because 1000 individuals was the sample goal, which was 28% of the population, 28% (or approximately) of each state's members were randomly chosen using a random numbers table.

every member of the sampling frame. With a systematic sample, every *n*th item in the sampling frame is included in the sample. Where to begin selecting the *n*th item is derived from a sampling interval established based on the ratio of the sample size to the population. A warning should be given when using systematic sampling. When sampling an organization with a rank or hierarchical structure one must be sure the selection procedure does not result in a rhythm with the organization's bureaucratic structure that would cause a particular type of individual to be selected each time. This negates the purpose of such a sample.

Cluster Samples

The last of the probability sampling methods is the cluster sample (also known as "area probability sample"). This sample consists of randomly selected groups, rather than individuals. The population to be surveyed is divided into clusters (e.g., census tracts). This is a multistage sample (sampling occurs two or more times) in which groups (clusters) are sampled initially (Adler & Clark, 2007). Subsequent subsamples of the clusters are then selected. For example, a sampling may be taken of correctional institutions nationwide. Employee information from each of the sampled institutions might then be obtained. From these lists, a sample of correctional employees from each institution may be drawn. This method is popular for national victimization or other national interest topics. It is a particularly useful tool for political scientists.

Usually, the researcher wants to make use of probability samples primarily because they are often more statistically stable. However, random sampling can be very expensive and logistically difficult to complete. Therefore, it is common to find nonprobability sampling in criminal justice and criminologic research.

Nonprobability Sampling

The major difference between probability and nonprobability sampling is that one provides the opportunity for all members of the sampling frame to be selected, whereas the other does not. This shortcoming of nonprobability sampling often leads to questions and concern over the representativeness of the sample. However, when the sample produces the requisite information, representativeness is often not as much of a concern (although its limitations must still be noted). Furthermore, a nonprobability sample could be perceived as representative if enough characteristics of the target population exist in the sample. There are four types of nonprobability samples: (1) purposive, (2) quota, (3) snowball, and (4) convenience.

Purposive Samples

Among the nonprobability samples, the purposive sample seems to be the most popular. Based on the researcher's skill, judgment, and needs, an appropriate sample is selected (Dunn, 2009). When the subjects are selected in advance based on the researcher's view that they reflect normal or average scores, this process is sometimes referred to as "typical-case sampling." If subgroups are sampled to permit comparisons among them, this technique is known as "stratified purposeful sampling" (Adler & Clark, 2007). A major factor of purposive sampling is accessibility to units or individuals that are part of the target population.

In all of the studies from the preceding **From the Real World,** the researchers chose their samples because they believed they best fit the needs of the study. The selection was based on their knowledge of the topic, the target populations, and accessibility. Although the samples may not have been representative, they did provide the requisite data to complete the studies. In most cases, because the purposive sample is not representative, findings cannot be generalized to complete populations. This does not mean that a purposive sample cannot be generalizable, but it must conform to some similar elements of the population. Despite statistical concerns about the sample, purposive sampling can offer researchers a legitimate and acceptable means of collecting data.

FROM THE REAL WORLD

Monto and Julka (2009) framing prostitution as an economic exchange, evaluated some of the consequences of conceiving of sex as a commodity rather than as an aspect of an intimate interpersonal relationship among the customers of prostitutes. Their data were collected from a purposive sample of 700 men arrested while trying to hire street prostitutes. DeMatteo, Marlowe, Festinger, and Arabia (2009) used a purposive sample of 28 individuals assigned to a drug treatment program to study what effect the program may have had on the participants' continued drug use. To gather opinions of police officers with many years of experience responding to emergency calls and pursuing fleeing suspects Schultz, Hudak, and Alpert (2009) used data collected from participants of In-Service Training in Emergency Vehicle Operations and Police Pursuits.

Quota Samples

Sometimes a study simply needs data collected from a set amount of participants fitting the sampling needs. These types of research efforts often rely on quota sampling. For this type of sample, the proportions are based on the researcher's judgment for inclusion. Does the individual or unit fit the needs of the survey? Selection continues until enough individuals have been chosen to fill out the sample. For example, assume one is required to conduct a study of high school drug use with a sample size of 100 first-semester college freshman at a given university. Because there is no means of identifying these individuals, a booth is set up in the student union where students are stopped as they come by and inquiry is made as to their status at the university. Only those who advise that they are first-semester freshman students are surveyed, and this approach is continued until the desired sample size or quota is reached. To ensure some level of representativeness by gender and race, the quota is set at a percentage equivalent to what the university claims to have in its population. For example, by gender, the university is 45% female and 55% male, whereas racially it is 43% white, 37% African American, and 20% other. Reaching some minimal level of representativeness requires one to continue selecting students until the quota sample seems to be comprised in similar fashion to the university. Obviously, this can be a painstakingly long method and does not guarantee representation.

Snowball Samples

Although not a highly promoted form of quantitative sampling, snowball sampling is commonly used as a qualitative technique. The snowball sample begins with a person or persons who provide names of other persons for the sample. This sample type is most often seen used in exploratory studies where an appropriate target population is not readily identifiable, making a sampling frame more difficult to select, but does not eliminate the existence of an identifiable sampling frame (Senese, 1997). Additionally, despite the issue of representation, snowball sampling requires the researcher to rely on the expertise of others to identify prospective units for the sample. Snowball sampling is frequently used in field research when the researcher must rely on introductions from group members to access other group members.

Obviously, the snowball method may lead to an elite sample that has no representative or generalizable attributes. Consequently, if the data address the research question, then these shortcomings are acceptable.

Convenience Sample

The last choice for a sample is the convenience sample or available subjects sample. Here there is no attempt to ensure any type of representativeness.

FROM THE REAL WORLD

Flanyak (1999), conducting a qualitative, exploratory research study, used both snowball and convenience sampling methods. Through this method she obtained 22 subjects for the study. Although her sampling frame consisted of lists of sociology and criminal justice department faculty members in several universities in Illinois, it was the result of initial interviews during a pilot study with respondents at her academic institution that other potential subjects were identified. Furthermore, other subjects were chosen from the faculty of her undergraduate institution to ensure a high response rate.

Usually this sample is a very abstract representation of the population or target frame. Units or individuals are chosen simply because they were in the right place at the right time.

Analysis of data from a convenience sample is extremely limited. The sample selected may or may not represent the population that is being studied. Therefore, generalizations that are made about the population cannot be considered to be valid (Adler & Clark, 2007). Because of this limitation, convenience samples are not useful for explanation or even for description beyond the sample surveyed. They are often useful as explorations on which future research may be based.

The quality and quantity of the data are dependent on the sampling technique. Statistical support is stronger for probability samples. However, there are times when nonprobability samples are fruitful. Regardless of which type is used, an important element of each is the sample size.

FROM THE REAL WORLD

To study officer opinions on police misconduct, Hunter (1999) developed a survey instrument based on the findings of prior research on police ethics and misconduct. Before administering this survey to a sample of several hundred officers in the southeastern United States, he sought feedback from a convenience sampling of currently serving officers. One group consisted of officers on a patrol shift from a mid-sized metropolitan police agency. A second group was composed of officers taking college courses at a regional college. A third group was made up of officers in their final phase of training at a regional police academy. Findings from this convenience sample not only permitted Hunter to refine the questionnaire, but he was also able to gain insights as to how officers felt about different forms of police misconduct and a variety of proposed solutions aimed at curbing such behaviors. The results were only indicative of this sample's opinions and could not be construed to be representative of the opinions of officers outside the sample. However, as an exploratory study it was quite useful.

Sample Size

The quality of a sample is largely dependent on its size (Banyard & Grayson, 2009; Creswell, 2008; Frankfort-Nachmias & Leon-Guerrero, 2008). The belief is that the larger the sample, the more likely the data will more truly reflect the population. An interesting aspect about sample size seems to be that there is no ideal set size. Usually, the sample size is the result of several elements: (1) how accurate must be the sample; (2) economic feasibility (how much does one have to spend); (3) the availability of requisite variables (including any subcategories); and (4) accessibility to the target population. Furthermore, confidence level is extremely important.

Confidence Levels

Deciding how large the sample should be requires an understanding of confidence intervals, which indicates a range of numbers (e.g., ±15). Because a sample is merely an estimated reflection of the target population, a confidence interval suggests the accuracy of the estimate with the smaller the confidence interval the more accurate the estimated sample. The estimated probability that a population parameter will fall within a given confidence interval is known as the "confidence level" (Adler & Clark, 2007). Levels of 0.05 (95%) and 0.01 (99%) are most readily acceptable in most research (Gillham, 2009; Shaughnessy, Zechmeister, & Zechmeister, 2008). To reduce sampling error the researcher desires a smaller confidence interval. To do so, he or she selects a smaller confidence level. For example, Dantzker (2010) sought to meet an alpha or confidence level of 0.05, which required a sample size between 385 and 625. To reach this, over 1300 individuals need to be sampled.

In social science research, confidence levels are very important, although they are rarely explained or identified. In many cases readers are expected to accept that samples are statistically acceptable, or an accurate estimate of the target population. This does not mean that the findings should be ignored. It simply means that their application must be more conservative and judicious.

Sampling Formulas

The key aspect to selecting appropriate confidence levels is the sample size. As previously suggested, the larger the sample the more accurate the estimate. Therefore, it is beneficial to know just how large a sample is required to attain the best confidence level. There are several mathematical formulas to assist in determining sample size. Unless one is a good mathematician, it is suggested that preexisting tables or computer statistical packages be used to make that determination. Should those means not be available, the following paragraphs demonstrate how such a formula may be used.

A Commonly Used Sampling Formula

In selecting a sample size one is seeking to draw a large enough number of observations from the target population to ensure that the sample accurately represents that population. As was discussed previously, the larger the sample size, the more likely that it is representative. However, because the costs in time and money of sampling large numbers are prohibitive, probability theory is relied on to estimate the proper sample size. One is guided in the selection by knowledge of acceptable sample sizes. Generally, in social science research, one seeks a sample size that 95 times out of 100 varies by 5% or less from the population. In some cases one may use less stringent requirements and in others one may wish to have (and be able to afford) a higher level of accuracy.

Assuming that one wishes to have a sample of a large population that is at the 95% confidence level and has a sample error within 5% of the population, one could use the following formula

$$n = \frac{(1.96)^2[p\,(1-p)]}{se}$$

where n = sample size needed,
p = assumed population variance,
se = standard error, and
1.96 represents a normal curve z score value at a confidence level of 95%.

For an error tolerance of 5% at the 95% confidence level one would use 0.05 as the sample error. The formula then becomes:

$$n = \frac{(1.96)^2[.5\,(1-0.5\,)]}{0.05}$$

$n = 384.16$ which rounds up to $n = 385$

Thus, a sample size of 385 provides a sample that had an error tolerance of +5% at the 95% confidence level. If one wished to change confidence levels or error tolerance, one would then adapt the formula. For example, if one wanted to ensure that the sample was representative, and had the money and time to do so, an error tolerance of +1% at the 99% confidence level would result in the following:

$$n = \frac{(4.2930175)^2[0.1\,(1-0.1)]}{01}$$

$n = 16586.99$ or $= 16{,}587$

A Sampling Size Selection Chart

Having read how the previous formula was used to obtain the desired sample size, you may decide that you have no desire to ever do so. To save you from such an exercise, a simple chart is included that makes sample size selection

TABLE 9-1

Sample Size Selection Chart

Error tolerance	Confidence levels	
(percent)	95%	99%
1	9604	16,587
2	2401	4147
3	1068	1843
4	601	1037
5	385	664

Source: Adapted from Cole, R. L.. *Introduction to Political Science and Policy Research.* St. Martin's Press, 1996.

much easier. Simply determine the confidence level and error tolerance desired in the survey sample and look it up in Table 9-1. The number indicated is how many observations from the study population are need to select randomly to have a representative sample.

Table 9-1 provides the sample size needed based on the error tolerance and confidence level desired. However, those are the numbers of observations needed in the sample. To ensure that those numbers are obtained, it is recommended that one always oversample by 20%. If this is not enough, one can always add more observations as long as they are randomly selected from the same population and any time differences do not affect responses. Remember, the sample must be selected randomly if it is to be representative of the population.

Summary

Gaining data from a complete population is usually impossible. In most cases, research data are best obtainable through a sample. Before a sample is chosen, identifying the sampling frame is necessary. An identifiable sampling frame leads to a decision as to what type of sample to select: probability or nonprobability. Probability samples include random, stratified random, strata, and cluster sampling. Purposive, quota, snowball, and accidental are forms of nonprobability sampling.

Regardless of the type of sampling, there is a question of sample size. Although no magical number exists, confidence levels provide a statistical means for establishing legitimacy of the sample. The smaller the confidence level, the more representative is the sample. As can be seen, careful thought should be given before choosing a sample type and a sample size.

METHODOLOGICAL QUERIES

1   Because the number of correctional officers available for the study is not overwhelming, sampling does not seem to be a likely issue. Explain why.

2   If you were to have to sample, describe to the sheriff what sampling method you would use and why.

3   The sheriff wants to know how the results might apply if less than half of those surveyed return completed questions. Explain the importance of sample size. Include confidence intervals, confidence levels, and sampling error in your discussion. Determine how many questionnaires would be necessary to obtain a sample with an error tolerance of +3 at the 95 percent confidence level if the number of officers available were 650, 525, and 400.

Data Collection

What You Should Know!

The previous chapter explained how to identify and select a group from whom data is to be collected. Chapter Eight introduced the questionnaire. This chapter examines a variety of methods for collecting data. After studying this chapter the reader should be able to do the following:

1. Identify the four primary data collection techniques.
2. Explain the strengths and weaknesses of mail surveys.
3. Describe the strengths and weaknesses of self-administered surveys.
4. Compare and contrast structured, unstructured, and in-depth interviews.
5. Compare and contrast face-to-face and telephone interviews.
6. Discuss the strengths and weaknesses of observational research.
7. Explain the strengths and weaknesses of archival research.
8. Describe the strengths and weaknesses of content analysis.
9. Compare and contrast the advantages of survey, interview, observational, and unobtrusive research.
10. Identify and explain the disadvantages of survey, interview, observational, and unobtrusive research.

One of the most crucial aspects of the research effort is the collection of the data. Improperly collected or incorrect data can delay or even cause the cancellation of the research effort. Therefore, before the researcher begins any type of data collection, the individual must be sure to choose the right data collection technique. Although one of the best means of collecting data is through the experimental design, this method is not very conducive to social science research. However, there are alternatives that are very effective and efficient, especially for the justician and criminologist. There are four primary data collection techniques available: (1) survey, (2) interview, (3) observation, and (4) unobtrusive means. Interviews have been broken out of survey research for discussion purposes in this chapter, but keep in mind that interviews are usually considered to be a component of survey research. These research methods have been discussed in detail in previous chapters. The focus in this chapter is on the issues involved in data collection using these strategies.

Survey Research

The most frequently used method for data collection is the survey, despite the fact that the research may be formed around invalid assumptions (Talarico, 1980). Although an excellent tool for gathering primary data, the survey is often misunderstood. It is quite useful in both descriptive and analytical studies. In criminal justice and criminology some of the uses include measuring attitudes, fears, perceptions, and victimizations. There are two primary means for collecting data through surveys: self-completing questionnaires and interviews. A common means of distributing questionnaires is through the mail or by direct distribution. Interviews can be conducted in person or by telephone.

Mail Surveys

Although there are many means available for conducting surveys, one of the most popular approaches is distributing surveys through the mail. This method allows for use of fairly large samples, broader area coverage, and minimized cost in terms of time and money. Additional advantages include that no field staff is required, the bias effect possible in interviews is eliminated, the respondents are allowed greater privacy, fewer time constraints are placed on the respondents so that more consideration can be given to the answers, and the chance of a high percentage of returns improves the representativeness of the sample.

The mail-survey method is extremely advantageous, but it has numerous disadvantages. One of the most frustrating disadvantages is the lack of response, or nonresponse. Therefore, it is common to send reminders to the participants. Although this helps increase the number of respondents, it also can add costs not part of the original research plan.

A second disadvantage is the possible differences that might exist between the respondents and nonrespondents. For example, what if only individuals who had an interest in the topic responded? The findings might then be biased toward a particular response, which may please some, but would not really provide valid and reliable information.

Still another disadvantage depends on the type of survey sent. A lack of uniformity in responses could present problems when open-ended questions are used. Because each respondent answers in a different manner (e.g., paragraph, listing, or single word), this could make data compiling more difficult. In addition to the lack of uniformity is the problem that misinterpretation of the question could also create glitches in the data. Finally, slow return rates can delay the project.

The disadvantages of mail surveys can be disconcerting, but they are not difficult to overcome. The key to this type of data collection is to attain the highest response rate possible. The follow-up is just one way. Other ways to increase response rates include the following (Adler & Clark, 2007; Bachman & Schutt, 2008; Dunn, 2009; Frankfort-Nachmias & Leon-Guerrero, 2008; Shaughnessy, Zechmeister, & Zechmeister, 2008):

1. Offering some type of remuneration or "reward" for completing the survey. One popular method is to offer a cash incentive (e.g., sending a dollar bill with the survey).

2. Appealing to the respondent's altruistic side by advising the perspective respondent that his or her response would be extremely helpful in learning more about the subject of the research.

3. Using an attractive and shortened format. Although this may sound silly, it is amazing how much more likely someone is to respond to a questionnaire that is eye-catching. This tends to show the respondent that some time and thought were given to this questionnaire and may be more inspiring to complete than designs that are very lackluster. Furthermore, keeping the format short and simple is more likely to generate a better response than a lengthy questionnaire.

4. Indicating that the survey is sponsored or endorsed by a recognizable entity. Respondents may be more encouraged to respond to a survey when they recognize and respect an entity supporting the research.

5. Personalizing the survey. Addressing the questionnaire to a specific person often adds more legitimacy to the research as opposed to receiving something that simply says "Dear occupant or resident."

6. The timing of the survey. When a survey is sent could be extremely important. For example, many individuals spent hours and hours watching the O.J. Simpson trial. Research geared toward public perceptions of the court process as observed in this trial probably would have received much better response immediately following the completion of the trial than if such a study were to be conducted today.

Acceptable Mail Survey Response Rates

The previous section discussed the frustration of dealing with low response rates and provided recommendations for enhancing return rates. Perhaps the best way

FROM THE REAL WORLD

To examine social determinants that explain correctional officer exposure to blood and bodily fluids, Alarid (2009) conducted a study of experienced correctional officers from five prisons. They were selected at random and sent an anonymous mail survey regarding situations that may have placed them at risk for exposure to HIV while at work. Although a total of 500 officers were selected, only 192 surveys were returned for a response rate of 38.4%. Of the 192, a total of 16 of the surveys were not included because of a disproportionate number of incomplete responses making the final sample size 176.

to cope with low return rates is (in addition to following the recommendations that were provided) to allow for them in the research design. In Chapter 9, it was recommend that one oversample to meet sample size requirements. In mail surveys the 20% oversampling rate may not be enough. Suggested response rates for mail surveys have been given as (1) 40% within 2 weeks, (2) 20% within 2 weeks of a follow-up letter, and (3) 10% within 2 weeks of the final contact (Adler & Clark, 2007; Dunn, 2009; Frankfort-Nachmias & Leon-Guerrero, 2008). Furthermore, response rates of 50% are adequate for analysis and reporting, 60% is good, and 70% is very good (Adler & Clark, 2007; Dunn, 2009). These numbers are consistent with our own experiences. Note that based on the return rate indicated previously, it may require several follow-ups to obtain a good or even adequate response rate.

The response rate varies depending on the type of survey and the targeted respondents. Ordinary citizens are far less likely to respond to a general survey than are members of a constituent group being polled by an organization in which they hold membership. In some situations, a 100% response rate may be possible. In all, a mail survey can be very effective and efficient.

Self-Administered Surveys

More and more, personal or telephone access to particular populations is becoming increasingly difficult for researchers. Some populations might need to be kept restricted (e.g., prison inmates), whereas others wish to remain anonymous or do not want to allow personal contact with researchers (e.g., police officers). However, these types of populations can offer vital data regarding numerous research topics. This is where the self-administered survey is applicable. The self-administered survey is generally a written questionnaire that is distributed to the selected sample in a structured environment. Respondents are allowed to complete the survey within a given time period and then return it to the researcher, often through an emissary of the sample.

Although self-administered survey data collection has long been a popular means of obtaining data from inaccessible populations, the advent of the

Internet has provided a growing means of data collection. With the Internet researchers can now reach target populations that are literally located throughout the world or simply within a single country or state. A popular Internet site for conducting surveys is SurveyMonkey, which began in 1999. SurveyMonkey.com allows the researcher choice of designs and question types, various collection methods and restrictions, and analytical features. Furthermore, it provides quick access for respondents and allows them to complete a survey at their convenience. Dantzker (2010) used this method to conduct his study among clinical psychologists in the United States by setting up the questionnaire on SurveyMonkey.com and sending the sample members a link to the questionnaire. This allowed the respondent to complete it at her or his leisure. The only constraint was on how long the questionnaire was available.

The advantages of the self-administered method include targeting large samples, covering wide geographic areas, cost efficiency, ease of data processing, and the ability to address a wide variety of topics. The disadvantages

FROM THE REAL WORLD

Hunter's (1988) research of convenience store robberies used mail surveys that were sent to approximately 130 law enforcement agencies seeking robbery data on the 200 stores in the sample. The letter requesting assistance was from the Florida Attorney General to the Sheriff or Chief of Police of the agency stating their mutual interest in combating robbery and requesting assistance. The letter from a high-ranking state official, the brevity of the survey, and the mutual interest in the survey topic resulted in an initial response of about 67%. A follow-up packet containing the original letter from the attorney general and a new survey was sent after one month (note with large bureaucracies, particularly if the data requested must be looked up or calculated, a one-month response rate is not unreasonable). The follow-up resulted in a return rate of 92% by the end of the second month.

A follow-up telephone call was then made to the 8% of agencies that had not responded. In the telephone calls respondents were asked either to respond by mail or by telephone in the next week. All but two agencies did so. Another follow-up telephone call to those agencies finally resulted in the information. The records manager in one agency stated that neither store in his jurisdiction had been robbed in the preceding three years. However, over a year later, after the research had been published, the records manager returned the forms indicating two robberies at one store and three at the other. A survey from a student or a professor without the connection with the Attorney General's Office would have been fortunate to have achieved a 60% response rate. In all, a mail survey can be very effective and efficient. Yet, there are situations where a mail survey is not feasible but the survey method is still necessary to complete the research. Self-administered surveys help fill this need.

FROM THE REAL WORLD

To examine the impact of a wide range of police stressors on potential health outcomes while controlling for various coping strategies in a large sample of urban police officers, Gershon, Barocas, Canton, Li, and Vlahov (2009) used a self-administered questionnaire. Their sample was recruited during roll calls at each of the department's nine districts and at three other major divisions, including headquarters. Officers completed the questionnaires before going out on their shift.

Lurigio, Greenleaf, and Flexon (2009) explored whether the police-related views of African American and Latino students differ with respect to three major predictive factors. Their survey data were obtained from students who were enrolled in 18 Chicago public schools. Student surveys were anonymous and information was not collected on the characteristics of the individual schools. The questionnaire consisting of 131 items in open- and closed-ended response formats was distributed during advising periods that the school had reserved for standardized test administration.

include a low return rate, nonresponse to some questions, and misinterpretation or misunderstanding of the questions. Disadvantages aside, self-administered surveys allow for greater diversity in criminal justice and criminologic research. Overall, self-administered surveys offer economic and efficient means of collecting data. Still, this type of survey does not typically allow for in-depth responses or for the researcher to follow-up on why a particular response was given. This is where interviewing is appropriate.

Interviews

For the purpose of this text, interviewing is viewed as the interaction between two individuals where one of the individual's goals is to obtain recognizable responses to specific questions. Interviewing is not an easy task and to be reasonably good at it often requires years of training. However, this does not preclude many researchers from trying this method. There are three types of possible interviews: (1) structured, (2) unstructured, and (3) in-depth. Each can be conducted in person or by telephone.

Structured Interviews

The most used type of interview in criminal justice research is the structured interview. This requires the use of closed-ended questions that every individual interviewed must be asked in the same order. Responses are set and can be checked off by the interviewer. The advantage to this type of interview is that it can be easily administered, has high response rates, and makes data processing much easier. It is in reality a questionnaire that is being administered orally with the interviewer completing the form for the respondent. Disadvantages include that it does not allow for further exploration of the responses attained, it is time-consuming and costly, and it can limit the types of responses given.

Unstructured Interviews

The unstructured interview offers respondents open-ended questions where no set response is provided. Although this allows for more in-depth responses, it is much more difficult to quantify the responses. It is also more susceptible to intervening or biasing elements. Unstructured interviews are often conducted in a field research. They require a very experienced and disciplined interviewer to be successful.

In-depth Interviews

Finally, the in-depth interview can make use of both fixed and open-ended questions. The difference from the other types is that the interviewer can further explore why the response was given and can ask additional qualifying questions. The advantage to this method of interviewing is that a substantial amount of information can be obtained. The disadvantages include that it is time-consuming, requires small samples, and can limit the topics researched. Like the unstructured interview, this technique is best left to experienced researchers who are very familiar with the issue being studied and how to conduct in-depth interviews.

Face-to-Face Interviews

In general, there are many advantages to using interviewing as a data collection tool. However, they differ slightly when the interview is completed face-to-face rather than by telephone. With face-to-face interviews, it provides contact between the researcher and the respondent. This contact can

FROM THE REAL WORLD

As previously cited, Dantzker (2010) created a questionnaire for his study. The questionnaire was initially derived from information obtained through a telephone interview with the target police agency's recruit division, applicant processing office, or the Chief's office. After identifying himself and the purpose of the call, if the agency contact person could not provide the information sought, contact information of the appropriate person was provided or a request was left for that person to return the call. Once contact was made with the appropriate person two basic questions of a structured interview were asked and responses recorded: What psychologic protocol(s) are used to screen police applicants, and what is the general content of the psychologic interview? (Dantzker & McCoy, 2006).

FROM THE REAL WORLD

To study how academic researchers gain access to data, Flanyak (1999) used in-depth interviewing. This method provided a high response rate, and allowed her to have personal contact with her subjects and to observe them during the interview. To assist in conformity, she made use of an interview guide instrument, which provided her with the opportunity and discretion to explore or probe respondents for detail in specific areas. Each interview lasted from 45 to 75 minutes, with the longest being two hours in length. Follow-up interviews were completed as needed and lasted approximately 15 minutes.

be positive reinforcement for participating in the research. Often simply receiving a questionnaire in the mail can be very sterile, and because of that lack of personal touch would-be respondents may simply ignore the survey. The interview also can usually guarantee a higher response rate.

Another advantage to this type of interview is that any misunderstandings or confusion can be cleared up. This helps ensure that the responses are truer to the question. It also allows for the researcher to act as an observer, giving the interviewer the opportunity to focus on nonverbal cues. The other advantages to the face-to-face interview include being able to use audiovisual aids, schedule additional interviews, make use of language the respondent can relate to, and discretion.

Although the advantages clearly outweigh the disadvantages, they cannot be ignored. There are four basic disadvantages to face-to-face interviews: (1) they are extremely time-consuming and costly, (2) there is interviewer effect or bias, (3) there is interviewer error, and (4) the interviewer's skills or lack of them could be detrimental (Bachman & Schutt, 2008; Dunn, 2009; Shaughnessy, Zechmeister, & Zechmeister, 2008).

It is common for researchers to make use of graduate students to assist in conducting interviews. Unfortunately, few graduate students come equipped with the skills required to conduct a good interview. Failing to provide or hone the necessary skills negatively affects the data. Therefore, to avoid interviewer errors it is important properly to train those who will be conducting the interview. Furthermore, interview results can be enhanced through the use of audio and video taping.

The face-to-face interview can be a very useful data collection technique. However, there are times when it is not possible to conduct face-to-face interviews, but the interview method is still necessary. This is when the telephone survey is useful.

Telephone Surveys

The advantages of telephone surveying begin with being able to eliminate a field staff and can allow for the creation of a very small in-house staff. It is also easier to monitor interviewer bias; specifically, it eliminates being able to send any nonverbal cues. Although not common, long in-depth interviews could be completed by the telephone. Finally, the telephone interview is less expensive and can be quick. The disadvantages to the telephone interview include (1) limiting of the scope of research, (2) difficulty in obtaining in-depth responses, (3) elimination from the sample parameters of anyone without a telephone, and (4) possible high refusal rates (Adler & Clark, 2007; Bachman & Schutt, 2008; Dunn, 2009; Shaughnessy, Zechmeister, & Zechmeister, 2008).

Comparison of Survey Strategies

There are two methods of collecting data through surveys: self-administered and interviews. Each method has submethods of collections, such as mail, Internet, face-to-face, and telephone interviewing. A comparison of similar criteria finds that there are advantages and disadvantages to each (Table 10-1). Ultimately, it is up to the researcher to determine which might best meet the needs of the research topic. Furthermore, it is possible to combine methods.

Even though combinations are possible, there are times when none of these methods may be practical, yet a survey technique may still be necessary to obtain the data. Survey data collecting, regardless of the means, is very

TABLE 10-1

Comparison of Survey Methods

Criteria	Mail	Internet	S/A	P/I	T/I
Cost	Low	Low	Low	High	Moderate
Response rates	Moderate	Moderate	High	High	High
Control	None	None	High	High	Moderate
Diverse population	High	Moderate	High	Moderate	Moderate
In-depth information	Low	Low	Moderate	High	Moderate
Timelines	Slow	Moderate	Fast	Slow	Fast

Abbreviations: S/A: self-administered; P/I: personal interview; T/I: telephone interview.

popular. However, it is not always the most appropriate and best means to collect data. Therefore, other ways of collecting data must be available, and there are several other methods.

Other Data Collection Techniques

Assume one wants to understand the inner workings of a gang or how particular patrol techniques are affecting citizen satisfaction. One could probably conduct interviews or create a self-administered survey, but neither of these may be able to give a complete picture because one may not know everything that needs to be asked. In some instances, observation is the best method for collecting data.

Observation/Field Research

Observation can fall into one of four approaches: (1) the full participant, (2) participant researcher, (3) the researcher who participates, and (4) the complete researcher (Senese, 1997). Regardless of which approach is used, observation allows the researcher to see firsthand how or why something works. It provides an opportunity to become aware of aspects unfamiliar to those who do not have firsthand experience.

To conduct observational data collection, the researcher needs to (1) decide where the observations are to be done, (2) decide on the focus of the observations, and (3) determine when the observations will be conducted (Senese, 1997; Shaughnessy, Zechmeister, & Zechmeister, 2006).

A choice of where observations are to be done may be quite limited. Although the idea is to determine whether the observations should take place in public or private, in many instances that choice does not exist. For example, to observe how incarcerated juveniles respond to a particular treatment program may only be possible in the institutional setting. However, if the program is provided to nonincarcerated juveniles, then the observations may take place in a public setting (e.g., school). Regardless of the choice, it is still an important one. The wrong setting generally means an unsuccessful research venture.

What is it that one wants to know? Deciding what aspect to observe is another important element. If one knows absolutely nothing about a phenomenon or entity, they may end up observing everything about it. The researcher needs to have an idea of what aspects are to be studied and how to do so. If the research is quantitative, a checklist needs to be prepared in advance rather than relying on field notes (Figure 10-1). Finally, the period or time frame when the observations are to be conducted should be determined to provide the best possible opportunity of collecting the desired data.

Store: _____

Location: _____

City (if applicable): _____

County: _____

Law Enforcement jurisdiction: _____

Variable Score:	1	2	3	
1. Location of cashier:	Cent.	Side		____
2. Number of clerks:	Three	Two	One	____
3. Visibility within:	Good	Poor		____
4. Visibility from outside:	Good	Poor		____
5. Within two blocks of major street:	No	Yes		____
6. Amount of vehicular traffic:	Heav.	Mod.	Light	____
7. Type of land use nearby:	Com.	Res.	Unused	____
8. Concealed access/escape:	No	Yes		____
9. Evening commercial activity nearby:	Yes	No		____
10. Exterior well lighted:	Yes	No		____
11. Gas pumps:	Yes	No		____
12. Security devices in use:	Yes	No		____
13. Cash handling procedures:	Good	Poor		____
14. Hours of operation:	Day	6–12	24 hr	____

Note that the checklist provides for either circling answers or edge coding as an aid in the later inputting of data. Also, the order of attributes (Yes No, No Yes) are reversed in some questions to ensure that the numerical coding 1, 2, or 3 is consistent with increasing vulnerability to robbery.

FIGURE 10-1 On-site evaluation form.
Source: Reproduced from R. D. Hunter. *The Effects of Environmental Factors Upon Convenience Store Robberies in Florida.* Florida Department of Legal Affairs, 1988. Courtesy of the Florida Department of Legal Affairs.

Like survey research, observational research has its advantages and disadvantages. One of the greatest advantages is the direct collection of the data. Rather than having to rely on what others have seen, here the researcher relies only on his or her observations. However, that could also serve as a disadvantage because of researcher misinterpretation or not understanding what is seen. An extremely important facet is recording the observations either in field notes, with an audio recorder, or by video as quickly and as accurately as possible. This also helps reduce inaccuracy and inconsistency. It is further recommended that one transcribe field notes as soon as possible afterward to ensure interpretation of hurried handwriting and any abbreviations that were used while they are still fresh in one's mind. The fact that the research is being conducted in the phenomenon's natural environment is a bonus. Recall that a shortcoming of experiments is that the environment is controlled, perhaps biasing the results.

Observational research takes place in the real world, legitimizing the observations. "This is what actually happened," rather than "This is what should happen." Observational data collection, too, may not be used in all circumstances. What happens when a topic deals with a phenomenon that has already occurred and can no longer be observed or surveyed? What if one wants to research a phenomenon without having to "disturb" anyone? This type of data collection would rely on using unobtrusive data.

Unobtrusive Data Collection

The last means for collecting data is through what are referred to as "unobtrusive measures." These are any methods of data collection where the subjects of the research are completely unaware of being studied and where that study is not observational. It is a method that prevents the researcher from direct interaction or involvement with what is being studied. Two of the more commonly used types of unobtrusive data collection are the use of archival data (analyzing existing statistics or documents) and content analysis.

Archival Data

In criminal justice and criminology using official statistics or records is viewed as a form of archival collection. For example, one of the most common means of studying crime is through the use of the Uniform Crime Reports (UCR) published yearly by the Federal Bureau of Investigation. Data in the UCR are collected monthly from police agencies, on a voluntary basis. Because the UCR provides crime data from all over the country, and includes arrest and offense information, a researcher could do an in-depth analysis of crime patterns or arrest characteristics without the arrestees or victims having any knowledge of the research in progress.

Another use of archival data is historical research. Rather than a quantitative analysis of existing statistics, historical analysis is generally a qualitative analysis of documents and prior research. Archival data is often found in two types of records: public and private. Public records include actuarial, political, judicial, and other governmental documents, and the mass media. Private records include diaries, letters, foundations, and autobiographies. One of the biggest problems with archival data is proving the authenticity of the data. Still, archival collection allows researchers to be less intrusive than other means, and usually offers sufficient data to assess the phenomenon even though it can no longer be observed. One of the benefits of archival data is that there are many possible sources for information. Examples of archival data sources can be found in Box 10-1. Regardless of what types of records are used, archival collection can offer the researcher a wide array of topics and data. The biggest problem to overcome is access. However, this can be done by making the right inquiries with time to wade through the bureaucratic processes.

Content Analysis

The other unobtrusive means is content analysis. This technique has been discussed in previous chapters and is a favorite strategy among researchers who wish to compare social events from different eras. Analysis of the contents of archival documents could be included in this research. Generally, the focus of the research is on publications and presentations (particularly media documents, such as newspapers, magazines, and television programs). Depending on how the research is structured, the analysis may be qualitative or quantitative in nature.

When engaging in content analysis, it is recommended that one try to create a quantifiable instrument to use. This requires serious thought about the issue being researched and the creation of a code sheet or checklist to record observations on a daily, weekly, or yearly basis (depending on the nature of the

---

Box 10-1

Example of Archival Data Sources

Records from various criminal justice agencies
    Arrest records
    Court dockets
    Prosecuting attorney's records
    Probation records
    Prison files
Data set sources
Uniform Crime Reports
National Victimization Survey
National Archive of Criminal Justice Data
National Criminal Justice Reference Service
United Nations Surveys of Crime Trends and Operations of CJ Systems
Bureau of Justice Statistics
NYS Division of Criminal Justice Services

TABLE 10-2

Survey	Interview	Observation	Unobtrusive Means
Advantages			
Wide variety of topics	Wide variety of topics	Actual observation of behaviors	Wide variety of topics
Simple to administer	Can give complex answers	Ability to link behavior to concept	Simplicity
Cost effective	Clarifications	Can be anonymous	Cost effective
Anonymous	Increase response rates		Anonymous
Comparable	Nonverbal cues		Comparable
Disadvantages			
Misinterpretation	Misinterpretation	Observer influence	Misinterpretation
Nonresponses	Interviewer biases	Current behaviors only	Researcher biases
Low return rates	Costly	Costly	Limited to availability
No clarifying		Unobservable behaviors	No clarifying
Limited answers			Limited answers

research). A summary sheet or form that permits the code sheets to be easily totaled is also recommended. These summary forms can then be used to input data for analysis. However, a checklist, in and of itself, is not as important as having a list of those items one wants to collect or observe.

There are a variety of means for collecting data. Ultimately, it is the task of the researcher to be able properly to choose one or more means that provide the best access to the required data (Table 10-2). Having provided the foundation of the concept of surveys, the next step is to understand and develop the primary survey instrument, the questionnaire.

Summary

Data collection can take many forms. The most popular method tends to be the survey. Surveys can be either self-administered or interviews. Self-administered surveys can be accomplished through the mail or by direct distribution to respondents. Interviews can be conducted in person (face-to-face) or over the telephone. They may be structured, unstructured, or in-depth. Other means of collecting data include observation, archival, and through unobtrusive means.

METHODOLOGICAL QUERIES

1   Despite the agreement that a survey will be the best method to use in the proposed study, explain to the sheriff the four primary data collection techniques that are available.

2   A self-administered survey will be what you'll use if you conduct the study. Describe the strengths and weaknesses of self-administered surveys.

3   Explain why none of the following would be useful for the proposed study:
    a Structured, unstructured, and in-depth interviews
    b Face-to-face and telephone interviews
    c Observation
    d Archival research
    e Content analysis

Final Steps

Data Processing and Analysis

What You Should Know!

At this point all the necessary steps have been established for the collection of the data. The next step is to establish the data in a manner from which analyses can occur. This chapter describes the process for preparing and analyzing data. After studying this chapter the reader should be able to do the following:

1. Identify and discuss the three stages of data processing.
2. Describe what occurs during the data coding process.
3. Describe the data-cleaning process.
4. Compare and contrast the different means of dealing with missing data.
5. Explain why data may need to be recoded. Provide an example of how recoding may be done.
6. Identify and describe the three types of data analysis.
7. Recognize and explain the three types of statistical analysis.
8. Present and discuss the four types of frequency distributions.
9. Describe the various means of presenting data in addition to frequency tables.
10. Identify and discuss the three measures of central tendency.
11. Present and discuss the three measures of variability.
12. Discuss what is meant by the terms "skewness" and "kurtosis."

To this point the reader has been indoctrinated with a wealth of information on how to go about conducting research. The time has finally come to address what to do with that data once they are collected: processing and analysis. A previous chapter advised that "data" is the plural of "datum." One must keep in mind that whenever referring to data in research, it should be used as a plural.

Be aware that this chapter is not intended to teach how to do statistical analyses; that is for another book and a different class. Instead, what this chapter provides is a general overview of the issues involved in processing and analyzing data and an introduction to the terminology needed to communicate with whoever might be analyzing the data.

Data Processing

Will the results support the hypothesis? Has something new been discovered? Are these findings consistent with previous research? These are examples of questions one hopes to answer through the data analyses. Thus, the data are collected and ready to be analyzed, now what? The first task is to prepare the data for analysis, what is generally referred to as "data processing." Data processing consists of data coding, data entry, and data cleaning.

Data Coding

Coding is simply assigning values to the data for statistical analyses. Not all data need to be coded. Quantitative data that are already in a numerical format can be left as is (i.e., age or numerical responses to scales). Nonnumerical variables (i.e., gender, marital status, or race), however, need to be coded. How these are coded is up to the researcher, but it is common to use a standard numbering format.

It is usually easier for the researcher if the coding scheme is developed in advance. "Precoding" is when the coding scheme has been incorporated into the questionnaire or observational checklist design. Such a design may enable the researcher to input data directly from the survey form rather than having to translate it to a code sheet. If the survey form has provided for edge coding, it is very easy to input data directly into the computer. Refer back to Figure 10-1, the field research checklist. Note how the form allows for edge coding, and how the data are run to have consistent responses—that is, they run in a logical format based on knowledge of the subject so that during statistical analysis, positive and negative influences are consistent in their directionality.

During the coding process qualitative data may also be converted into quantitative data. For example, an open-ended question about prior arrests may later be converted into numerical groupings based on the number of arrests, type of offense, seriousness of the charges, conviction versus acquittal, or other logical groupings created by the researcher. Remember the previous warning about false precision when doing such a conversion. Be certain that there is an explainable logic to the numerical assignments.

Ultimately, it is the researcher's decision to code the data in a manner that is best for analyses. Although part of this depends on the computer statistics package used, the coding should be clear. To ensure clarity create codebooks or coding keys to show others exactly where the variables and attributes are located. The coding scheme for a survey document is often a one-page document referred to as the "codebook." Figure 11-1 is an excerpt from Dantzker's (2010) study of police preemployment psychologic screening.

Data Entry

Once the data are coded they can then be entered into a computer software program for analyses. Figure 11-2 is an excerpt of the coded data after they have been entered into a statistical program for analysis.

Coded-Data Example, Computer Entries

At various points in the text, it has been suggested that a computer can assist in conducting differing aspects of the research process. The same is true for the analysis portion. Today there are a number of statistical software packages available to researchers, such as Excel, QuattroPro, Statistical Package for the Social Sciences (SPSS), and Statistical Analysis System. Each has its own quirks and specialties that require individuals to choose what best suits their needs and abilities.

Regardless of which is chosen, each allows for the entry, analyses, and storage of data. The authors recommend SPSS, which is available for personal computers after years of only being accessible through mainframes. It provides the ability to conduct almost every type of statistical technique required. It seems to be a common tool for many criminal justice and criminologic researchers. Before choosing any of the packages, the researcher should test several to see which is most comfortable to him or her. Once the choice is made it is simply a matter of learning how to use it and being able to interpret the results. Most statistical packages not only have handbooks but also excellent tutorials.

The key to data entry is accuracy. If possible, have a reliable person read the data aloud as it is entered into the computer to avoid having continuously to switch one's viewing from the code sheet to the computer screen. This helps prevent mistakes. However, one should still review what has been inputted to make certain that it is accurate.

If the data have been obtained from existing sources, such as a computer disk, and they are compatible with the statistical package being used, a great

Gender	Male	**1**
	Female	**2**

Years in practice as a clinical psychologist:

	< 1	**1**
	1–5	**2**
	6–10	**3**
	> 10	**4**

Association Membership:	APA	**1**
	IACP	**2**
	Both	**3**

Does your state mandate psychologic assessments for individuals being considered for employment with a police agency?

	Yes	**1**
	No	**2**
	I don't know	**3**

If "Yes": Does your state mandate the use of a particular assessment tool or tools?

	Yes	**1**
	No	**2**
	I don't know	**3**

Do you personally provide psychological screening services to any type of police/law enforcement agency?

	Yes	**1**

IF YES, please respond to the remainder of survey.

	No	**2**

IF NO, you may stop here. Thank you.

How do you identify or label yourself with respect to providing police screening services?

1   Police Psychologist (full time in-house psychologist)
2   Psychologist Consultant (full-time consultant to law enforcement)
3   Clinical Psychologist (occasional service provider to law enforcement)
4   Other

Years conducting police pre-employment screenings

	< 1	**1**
	1–5	**2**
	6–10	**3**
	> 10	**4**

Figure 11-1 Code book excerpt from prescreening of police/law enforcement candidates psychologist survey.
Source: Reproduced from Dantzker, M. L. *Psychologists' Role and Police Pre-Employment Psychological Screening.* ProQuest Company, 2010.

2	4	2	1	2	3	4
1	4	3	1	2	3	2
2	4	3	2		2	4
2	4	3	1	2	1	4
2	4	2	2	2	3	4
2	4	1	3		3	2
2	4	3	1	2	1	4
1	4	3	1	2	1	2
1	3	3	2		2	4
2	4	3	1	2	3	2
1	4	3	1	1	3	4
2	4	1	3	3	3	4
1	4	2	1	2	2	3
2	4	3	1	1	2	4
2	4	3	1	2	3	4
2	4	1	2		3	4
1	4	3	1		2	3
1	4	3	1	2	1	3
1	4	3	1	1	3	4
2	3	3	1	2	2	4
2	4	3	1	1	2	4
2	4	1	1	2	3	4
2	1	1	2		2	1
2	4	3	1	2	2	3
2	4	3	1	2	3	4
2	2	3	2		2	2
2	4	3	2		1	3
2	4	3	1	2	2	4
1	4	1	2		3	4
2	4	3	2		1	4
2	4	1	1	2	3	3
2	4	1	1	1	3	4
2	4	3	1	2	2	4
2	4	3	1	2	1	4
2	4	3	1	2	2	4
2	4	1	2	2	3	4
2	4	2	2		3	2
1	3	1	2		2	3
2	4	1	2	2	2	4

FIGURE 11-2 Coded data excerpt for Dantzker (2010).
Source: Reproduced from Dantzker, M. L. *Psychologists' Role and Police Pre-Employment Psychological Screening.*
ProQuest Company, 2010.

deal of time is saved. If a telephone survey was conducted, it may have been possible to enter the data directly into the computer as the questions were asked. Another shortcut is to use optica-scan sheets, such as the Scantron sheets used in taking multiple-choice examinations. This permits the data to be entered directly from the sheets marked by the respondents. Regardless of which technique is used, the data need to be cleaned.

Data Cleaning

Data cleaning is the preliminary analysis of the data (Adler & Clark, 2007; Frankfort-Nachmias & Leon-Guerrero, 2008; Shaughnessy, Zechmeister, & Zechmeister, 2008). Any mistakes that might have occurred during the initial recording of data or data entry are "cleaned." The first step is reviewing what has been entered for accuracy.

If one has a computer program that is programmed to check for errors, it will either beep or refuse to accept data that does not meet the coding requirements for that variable. This is automatic data cleaning. For example, assume that a Likert scale was used where 1 = strongly agree, 2 = agree, 3 = neither agree nor disagree, 4 = disagree, and 5 = strongly disagree. If one tries to enter a 6 as a value, the program beeps to alert of the error.

Unfortunately, not all data-analysis programs have this automatic function. Therefore, cleaning must be done semimanually (e.g., using SPSS); that is, a frequency analysis is run on all the data. From the results, any incorrectly inputted value should be observable.

Another type of data cleaning is contingency cleaning. In this technique, review of the data is expedited by the knowledge that certain responses should only have been made by certain individuals. Not all females would indicate having given birth to a child, but no males should have indicated such.

As one is reviewing the data, there are certain questions that logically lead to similar responses on following questions. If they do not, that particular questionnaire or form should be reviewed to check the accuracy of what was entered. The next step is determining what to do with missing data.

Missing Data

If a great deal of data has been collected, despite best efforts, some information will be "lost." This may be caused by an oversight in data entry (which can usually be easily corrected), or it may be because the respondents or recorders accidentally overlooked or deliberately chose not to answer a particular question. If missing data is the product of a data-entry error, this mistake may easily be corrected by obtaining the right information from the survey form or code sheet. If it is caused by oversight or intentional omission, there are a number of ways to deal with this missing information.

If the data have been left out on a single question one may chose to input it as a nonresponse. If not using 0 as a response, 0 may be assigned as a value. Researchers frequently use 99 (assuming they are not using continuous variables in which 99 could be a response) to indicate missing data. This is the preferred method of dealing with nonresponses.

Another option for dealing with nonresponses is to assume that the missing data are caused by an oversight rather than an intentional omission. In this situation one might look at the other responses to try to determine what the missing response would most likely have been. For example, if on related questions a respondent had indicated support for strict law enforcement, one might assume a similar answer on the unanswered question. Although some researchers use this solution, it is not recommended because it can lead to challenges about the objectivity and validity of the analysis.

Yet another option is to exclude from the analysis the survey instrument containing the omission. If there are several nonresponses, that option may be appropriate. However, if only one or two questions in a survey are not answered, this solution can lead to the loss of worthwhile information.

If the instrument used is a Likert scale, the solution is simple. Classify the response as a "neither agree nor disagree" or "do not know" (depending on how the scale is worded). This solution allows the data to be used without the fear of skewing the results.

Recoding Data

Although some data may be received in a numerical format (e.g., age), sometimes for purposes of analysis these data are recoded to fit into groups. If such data as income or age has been gathered, one may wish to use a simpler method of analysis, such as cross-tabulations, to examine their relationships. If a frequencies table (discussed later in this chapter) is run, one would note that these categories are quite extensive. A comparison of age by income could result in a table that is lengthy and confusing. Assuming that age was presented in rows, and income in columns, if the people surveyed ranged from 20 years of age up to 70 years of age, and if their incomes ranged from $10,000 to $100,000, the resulting table would be enormous. Because each individual year of age would have to be represented, there would be 51 rows in this hypothetical table. Because there would likely be as many different incomes as there were respondents, one could end up with thousands of columns. Yet, by collapsing the categories into logical groupings, the data can be presented clearly and concisely (Table 11-1). The resulting table could easily be collapsed even further if necessary. In collapsing data recall the earlier discussion of the different levels of measurement. Higher-level data (e.g., ratio or interval data) can be collapsed down into lower-level data (e.g., ordinal or nominal data) but one cannot do the opposite and convert lower-level data to a higher level.

TABLE 11-1

Hypothetical Age by Income Comparison

(Ordinal Data) Age Income	20–29	30–39	40–49	50–59	60–69	70+
$10,000 to $19,000	50	30	20	10	05	15
$20,000 to $29,999	100	80	50	30	50	80
$30,000 to $39,999	80	100	80	60	80	60
$40,000 to $49,000	50	100	160	120	100	80
$50,000 to $59,000	30	80	120	150	120	50
$60,000 to $69,000	20	50	100	120	100	40
$70,000 to $79,999	10	40	80	100	80	30
$80,000 to $89,000	05	30	60	80	60	20
$90,000 to $99,000	05	20	40	60	40	10
$100,000+	00	10	30	50	30	10

Data Analysis

Now that the data have been entered, analysis of it may begin. There are three types of data analysis: (1) univariate, (2) bivariate, and (3) multivariate. With univariate analysis, an examination of the case distribution is conducted one variable at a time (Frankfort-Nachmias & Leon-Guerrero, 2008; Gavin, 2008; Walker & Maddan, 2009). The process by which this examination takes place is referred to as "descriptive statistics." Descriptive statistics provide an understanding of the variable being investigated. Descriptive statistics are discussed in detail later in this chapter.

Bivariate analysis is when the relationship between two variables is examined. This examination may be comparative or inferential depending on the nature of the analysis. If one is exploring the relationship between the two variables, this denotes descriptive statistics, but because it also compares the two variables, the process becomes comparative statistics. Comparative statistics usually involve analysis of the attributes of the variables being examined to describe better the relationship between the two variables. Comparative statistics is categorized as a separate category here, according to the authors' practice. Most methods texts include comparative in their discussion of descriptive statistics. Comparative statistics exist in a gray area between descriptive and inferential statistics. How one classifies them is not as important here as understanding what is involved.

If the hypothetical Table 11-1 were real it would be an example of comparative statistics. Comparative statistics are discussed further in bivariate techniques in Chapter 12. An example is provided near the end of this chapter.

If one of the variables is identified as being dependent and the other variable is identified as being independent, the analysis of their relationship becomes inferential. Inferential statistics simply mean that one is trying not only to describe the relationship, but also to use that knowledge to make predictions or inferences about the dependent variable based on the influence of the independent variable. Bivariate analysis may be descriptive, comparative, or inferential in nature.

The final type of statistical analysis is multivariate analysis. Multivariate analysis is the examination of three or more variables. This technique is inferential in nature in that one has usually already conducted both descriptive and comparative statistical analyses (through univariate and bivariate analyses) of the data and are now seeking to examine the relationships among several variables. From this examination, one tries to develop explanations for the observed relationships. Inferential statistics are covered in detail in Chapter 12.

Statistical Analysis

Statistics are data presented in a manner that best represents what it is the researcher wants to present. Statistical analyses might best be viewed as processes for problem-solving (Frankfort-Nachmias & Leon-Guerrero, 2008; Gavin, 2008; Walker & Maddan, 2009). Statistics are used in criminal justice and criminology to help describe a variety of associated aspects, such as crime rates, number of police officers, or prison populations. In addition, statistics can be used to make inferences about a phenomenon. As discussed in the preceding section, there are three types of statistics: (1) descriptive, whose function is describing the data; (2) comparative, whose function is to compare the attributes (or subgroups) of two variables; and (3) inferential, whose function is to make an inference, estimation, or prediction about the data.

The remainder of this chapter focuses on providing an overview of the processes and terminology of descriptive statistics. Understanding the basic principles is important so that one can comprehend the work of others and communicate with those who might be aiding with statistical analysis of the work.

Descriptive Statistics

When the researcher is interested in knowing selected characteristics about the sample, it requires some form of descriptive statistics. One of the most common descriptive statistics is the use of frequencies. When data are first collected they are referred to as "raw data"; that is, they are not neatly organized.

FROM THE REAL WORLD

The descriptive data in Table 11-2 come from Dantzker's (2010) study. These data describe the sample's characteristics of interest.

A first step to organizing the data, after coding and entry, is through a frequency distribution, which simply indicates the number of times a particular score or characteristic occurs in the sample. This can be reported in whole numbers and percentages. Frequencies are commonly used to describe sample characteristics.

Frequency Distributions

There are four types of frequency distributions: (1) absolute, (2) relative, (3) cumulative, and (4) cumulative relative (Frankfort-Nachmias & Leon-Guerrero, 2008;

TABLE 11-2

From the Real World: Sample Demographics

	Police psychologists	Clinical psychologists	Total
Gender			
Female	11 (31)[a] (50)[b]	11 (28) (50)	22 (29) (100)
Male	25 (69) (47)	28 (72) (53)	53 (71) (100)
Total	**36 (100) (48)**	**39 (100) (52)**	**75 (100) (100)**
Years providing police services			
< 1 years	1 (2) (33)	2 (5) (67)	3 (4) (100)
1–5 years	6 (16) (40)	9 (23) (60)	15 (20) (100)
6–10 years	10 (27) (76)	3 (7) (24)	13 (17) (100)
> 10 years	20 (55) (44)	25 (65) (56)	45 (59) (100)
Total	**37 (100) (47)**	**39 (100) (53)**	**76 (100) (100)**
Agency types served			
Municipal	5 (14) (29)	12 (32) (71)	17 (23) (100)
County	0	7 (18) (100)	7 (9) (100)
State	1 (3) (25)	3 (8) (75)	4 (5) (100)
Combo	31 (83) (66)	16 (42) (34)	47 (63) (100)
Total	**37 (100) (49)**	**38 (100) (51)**	**75 (100) (100)**

[a]Column Percentage   [b]Row Percentage

Source: Reproduced from Dantzker, M. L. *Psychologists' Role and Police Pre-Employment Psychological Screening.* ProQuest Company, 2010.

Gavin, 2008; Walker & Maddan, 2009). Each offers a statistically sound indication of the sample's composition. None is recommended over the others. Instead, it is suggested that the one that seems to be the most desirable for reporting the data contained in the research be selected.

Absolute frequency distributions simply display the data based on the numbers assigned. Relative frequency distributions are the percentage equivalent of absolute frequency distributions. When dealing with large numbers, it is usually much easier for the reader to interpret percentages than raw numbers. Cumulative frequency distributions enable the reader to see what the products of the grouping are in the frequencies table by adding the absolute frequency of each previous variable. Cumulative relative frequency distributions do the same as cumulative frequency distributions but add relative frequencies rather than numbers. Table 11-3 demonstrates how the four types of frequencies are related.

Displaying Frequencies

Frequencies, by whole numbers or percentages in table form, are just one means of describing the data. Other means include pie charts, bar graphs, histograms and polygons, line charts, and maps. As demonstrated in Figure 11-3 (a pie chart), the frequencies percentages can be pictorially depicted simply. Such a display is clear and easily interpreted. Bar charts are also

TABLE 11-3

Absolute, Relative, Cumulative, and Cumulative Relative Frequency Distributions of Targeted Sample of Psychologist Members of APA by Region

	Absolute frequency	Relative frequency	Cumulative frequency	Cumulative relative frequency
EN Central	531	14.7	531	14.7
ES Central	176	4.9	707	19.6
Mid Atlantic	672	18.6	1379	38.2
Mountain	210	5.8	1589	44
New England	351	9.7	1940	53.8
Pacific	607	16.8	2547	70.6
South Atlantic	650	18	3197	88.6
WN Central	221	6.1	3418	94.7
WS Central	191	5.3	3609	100
TOTAL	3609	100		

Source: Reproduced from Dantzker, M. L. *Psychologists' Role and Police Pre-Employment Psychological Screening.* ProQuest Company, 2010.

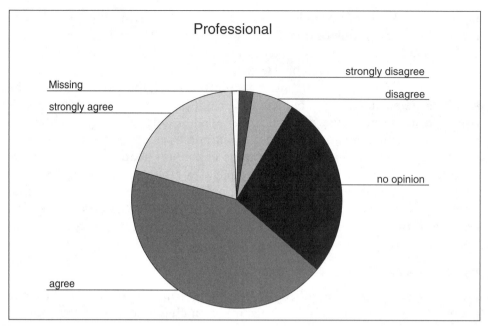

FIGURE 11-3 Pie chart.

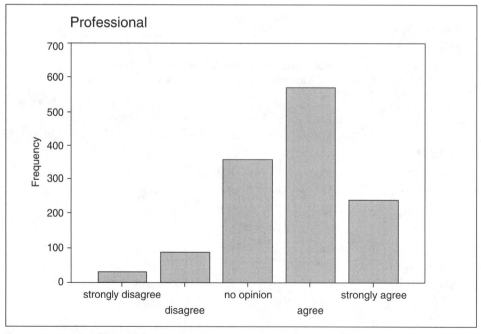

FIGURE 11-4 Bar chart.

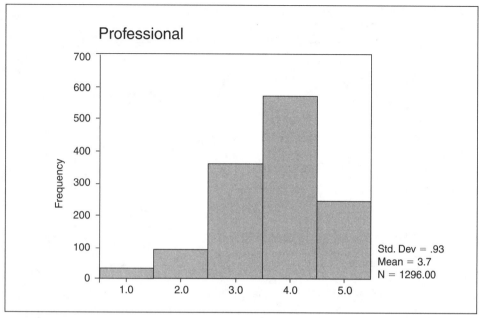

FIGURE 11-5 Histogram.

easily interpreted. They are used to display nominal- and ordinal-level data. Figure 11-4 demonstrates a bar chart. Histograms are bar graphs that are used to display interval and ratio data. They indicate the continuous nature of the variable by drawing the lines adjacent to each other. Figure 11-5 displays a histogram. Polygons display the same data as histograms but use dots instead of bars. Lines are then drawn between the dots to reveal the shape of the distribution. Figure 11-6 is an example of a frequency polygon.

In addition to these frequency-presentation techniques, criminologic researchers may also use line charts and maps. Line charts are polygons that demonstrate scores across time. As such, they are useful to reflect changes over time in the same dot-and-line format as frequency polygons. Figure 11-7 demonstrates a line chart.

In addition to the tables, graphs, and charts discussed previously, there are four other ways to describe the properties of the data: (1) measures of central tendency, (2) measures of variability, (3) skewness, and (4) kurtosis.

Measures of Central Tendency

Because frequencies can be quite cumbersome, researchers sometimes require a way to summarize the data in a simpler manner. Measures of central tendency are one way to summarize data. The three most common measures are the (1) mean, (2) median, and (3) mode. The mean is the arithmetic average. The median is the midpoint or the number that falls in the middle. The mode

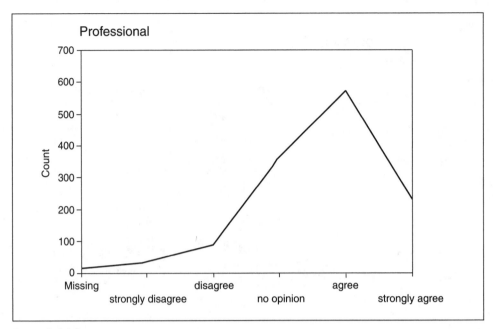

Figure 11-6 Polygon.

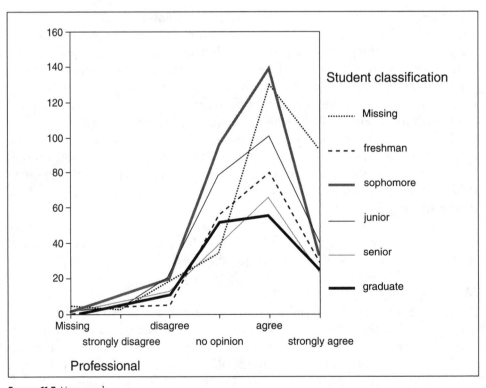

Figure 11-7 Line graph.

Sample data *n* = 100

Years of College Completed	Number of Responses	Responses x Years of College
1	10	10
2	20	40
3	30	90
4	20	80
5	10	50
6	8	48
7	2	14
	100	**332**

Mean = average = 332/100 = 3.32 years of college
Mode = most frequent (30 responses) = 3 years of college
Median = midpoint (30 responses are below 3 years, 40 responses above 3 years, 30 responses are within the 3 years grouping. The midpoint would be between the 20th and 21st respondents in the 3-year group.) = 3 years of college.

FIGURE 11-8 Measures of central tendency.

is the number that occurs most frequently. The mean is usually used as a measure of central tendency for interval- or ratio-level data. The median is used mostly for ordinal level data. The mode is generally used for nominal level data. Figure 11-8 shows how the three measures are obtained.

Using a measure of central tendency allows the researcher to simplify the numbers in a summary manner. Despite the more simplistic nature of these measures, researchers should be familiar with the characteristics of these measures before using them. In addition, these measures are often used in conjunction with measures of variability.

Measures of Variability

Despite their similarities, all statistics are not the same. The difference that occurs among statistics is called "variability." The three main measures are (1) range, (2) variance, and (3) standard deviation. The range is simply the difference between the highest and lowest scores. Variance is the difference between the scores and the mean. Standard deviation indicates how far the score actually is from the mean. To compute measures of dispersion, such as standard deviation, statisticians use z-scores. This text goes no further into the use of this methodology. If there is a need to calculate dispersion (variability) in one's research, it is recommended that one either take a course in statistics or consult with a statistician.

Skewness

The discussion thus far has alluded to the distribution of the data and has depicted it somewhat in the figures that were displayed. Many of the statistical techniques discussed in this chapter (and which will influence the discussions of inferential statistics in Chapter 12) are based on the assumption of a normal distribution. A normal distribution may also be referred to as a "normal," or "bell," curve. By using polygons or line charts or scatterplots (discussed in Chapter 12), researchers are able to see the distribution of the data. If it is a normal distribution, the researcher is able to use a broader range of statistical techniques. If it is a nonnormal (also known as "nonparametric") distribution, the statistical techniques that may be used are more limited.

The measures discussed previously (central tendency and variability) may be noted in the location of the center of the curve (central tendency) and in the spread of the curve (variability). They may also differ based on the symmetry of the curve. If one side has more values so as to cause the slope of that side to tail further outward, the distribution is said to be skewed to that side. Skewness alerts the researcher to the presence of outliers. Such knowledge aids the statistician in conducting analysis of the data.

Kurtosis

Kurtosis is another distribution consideration that warrants some discussion. Kurtosis refers to the amount of smoothness or pointedness of the curve. A high thin curve is described as being leptokurtic. A low curve is platykurtic. For the purposes of this text, one need only recognize the terms if a statistician states that the data distribution exhibits such features.

These statistics, in tandem with measures of central tendency and variability, provide useful ways to describe the data. In general, descriptive statistics are primarily used to describe how the data are comprised. However, describing the data is usually just a small portion of making inferences from it.

Comparative Statistics

As mentioned previously, descriptive statistics usually consist of univariate statistics that describe the data. Inferential statistics consist of either bivariate or multivariate analyses of the relationships among variables. Between the two lies comparative statistics. Comparative statistics are often classified as being part of descriptive statistics because they tend to provide more insight about the nature of the variables than their relationships with other variables. Depiction of crime rates over time, noting the percentage of change in criminal activity, and trend analyses are examples of comparative statistics. Depending on the text that is used, such activities may be categorized as either descriptive or inferential statistics. This text has discussed comparison as being within both.

Summary

Once the data collection is accomplished, the next two phases in completing the research are data processing and data analysis. Because the data are often obtained in a raw manner, they must be coded for entry into a statistical software program. Once the data have been entered they must be cleaned to ensure accuracy and appropriateness for subsequent statistical analysis. At this point missing data are dealt with and recoding may also be required.

Univariate analysis is conducted to gain knowledge about the data before any bivariate and multivariate analyses. Descriptive statistics are then used. Description may involve the use of frequency distribution tables and charts and graphs. The researcher also examines the data by using measures of central tendency and measures of variability. The data may be further assessed by reviewing the shape of the data distribution.

While the data are being examined the analysis may progress from description to comparison of variables. This process links description to inferential statistics, which are discussed in the next chapter.

METHODOLOGICAL QUERIES

1   The sheriff asks, "What will you do once you have collected the data?" Respond by identifying and discussing the three stages of data processing.
2   He then asks, "Is there a difference in the types of data? How you have to analyze them?" In response, identify and describe the three types of data analysis. Then add an explanation of the three types of statistical analysis.
3   Once the data is analyzed, how do your report or display it? Discuss the four types of frequency distributions, the various means of presenting data in addition to frequency tables.

Inferential Statistics

What You Should Know!

The task of inferential statistics is to allow the explanation of relationships and prediction of future events based on that knowledge. The type of inferential statistic used depends on the level of measurement used to gather the data. After studying this chapter the reader should be able to do the following:

1. Differentiate between descriptive, comparative, and inferential statistics.
2. Discuss measures of association and provide examples.
3. Explain what is meant by statistical significance and describe how tests of significance are used.
4. Present and describe the commonly used comparative statistics techniques.
5. Discuss bivariate analysis and provide examples.
6. Discuss multivariate statistics and provide an example of a multivariety technique for nominal and ordinal level data.
7. Describe the various multivariate techniques for interval level data.
8. Explain what is meant by nonparametric techniques.

Statistical Analysis

The previous chapter discussed the three types of statistics. Descriptive statistics describe the data being analyzed. A variety of descriptive statistics were presented in Chapter 11. Comparative statistics, whose function is to compare the attributes of two variables, were also introduced. Comparative statistics are within a "gray area" that moves from description to inference. As such, comparative statistics are often classified as either descriptive or inferential rather than as a separate category. An example of how comparative statistics might be used for description was provided.

Inferential statistics were also defined in Chapter 11. The purpose of inferential statistics is to make inferences, estimations, or predictions about the data. This chapter provides an overview of inferential statistics. How comparative statistics may be used to begin making inferences about the data is demonstrated. A variety of inferential techniques used by criminologic researchers is then presented and discussed.

Overview of Inferential Statistics

Inferential statistics allow the researcher to develop inferences (predictions) about the data. If the sample is representative, these predictions may be extended to the population from which the data were drawn. Inferential statistics allow criminologic researchers to conduct research that can be generalized to larger populations within society. When making inferences about data sets, researchers rely on measures of association to determine the strength of the relationships between or among variables. Tests of significance are also used to discover whether the sample that was examined is representative of the population from which it was drawn.

Measures of Association

Measures of association are the means by which researchers determine the strengths of relationships among the variables that are being studied. The measure of association that is used is dependent on the type of analysis being conducted, the distribution of the data, and the level of data under analysis. There are many measures of association used in criminologic research. Although individual researchers may prefer to use different measures, some measures are considered to be standards. Lambda is commonly used for nominal level data. Gamma is commonly used for ordinal level research. Pearson r and R^2 are commonly used for interval and ratio level data.

Typically, measures of association for nominal level data tell researchers how strong the relationship is but do not indicate directionality. A measure such as lambda runs from 0 (no relationship) to 1 (perfect relationship). A 0.00001 indicates a very weak relationship. A 0.9999 indicates an extremely strong relationship. Depending on the measure used (some tend to be more powerful than others) the researcher assesses the relationship as being weak, moderate, or strong. For example, a lambda score of 0.5367 indicates a moderate relationship, whereas a lambda score of 0.8249 is an indicator of a relatively strong relationship. The researcher has to assess the contingency table (cross-tabulation) to determine whether the relationship is positive or negative.

Measures of association for ordinal level data tell the researcher both the strength and the direction of the relationship. The ordinal measures range from −1 to +1. A zero (0) indicates no relationship; a negative one (−1) indicates a perfect negative relationship (as one variable increases, the other variable decreases); and a positive one (+1) indicates a perfect positive relationship (as one variable increases, the other variable increases). A score of −0.8790 for the commonly used gamma indicates a strong negative relationship, whereas a score of +0.2358 indicates a weak positive relationship. It should be noted that there are several other measures available for researchers to use (some of which have the advantage of being preproportional reduction in error measures that are more attractive to statisticians). The more commonly used measures are provided here. To review the advantages of other choices the reader should consult with a statistician, take a statistics course, or consult a statistics text.

Measures of association for interval and ratio data are determined by the statistical technique used. For correlation, Pearson r is the measure of association; for simple linear regression and multiple regression, R^2 is used. Other techniques use different measures. However, they like ordinal measures range from −1 to +1. Different interval measures are discussed within the various strategies presented.

Statistical Significance

Statistical significance is how researchers determine whether their sample findings are representative of the population they are studying. If one is using a complete enumeration (the entire population is studied rather than just a sample from that population) then statistical significance is a moot point in that it is already known that the population is accurately represented. However, complete enumerations are rare because very small populations limit generalizabilty and large populations are too costly and time-consuming to study as a whole.

Statistical significance is based on probability sampling. Nomothetic explanations of causality are based on probabilities. It is this use of probability that enables one to make inferences based on a relatively few observations. Generally, social science research desires a statistical significance of 0.05 or better. This means that one is 95% confident that the findings represent the population that was sampled. Before computers were readily accessible to researchers, the statistical significance of data was obtained by using calculators laboriously to figure mathematical equations. Today, thanks to statistical packages, this work is performed by computers, which provide the significance in exacting detail.

Comparative Statistics

There are several comparative techniques available to criminal justice researchers. The ones briefly discussed here are crime rates, crime-specific rates, percentage change, and trend analyses.

Crime Rates

Crime rates are perhaps the most frequently presented data within criminal justice and criminologic research. These are nothing more than the number of crimes for an area (city, county, state, and so forth) divided by the population for that area and then multiplied by 100,000. Crime rates for index crimes are commonly displayed in Uniform Crime Reports. Table 12-1 provides an example of crime rates. By viewing the rates presented in that table, the reader may easily compare the rates among the various states.

Crime-Specific Rates

Crime-specific rates differ from crime rates in that they use a different base than population within the computations. For example, one could look at motor

---

TABLE 12-1

Violent Crime Rates Per 100,000 Population for Select States, 2007

State	Population	Violent crime	Murder and nonnegligent manslaughter	Forcible rape	Robbery	Aggravated assault
Arizona	6,338,755	482.7	7.4	29.3	151,7	294.3
California	36,553,215	522.6	6.2	24.7	193	298.8
Hawaii	1,282,388	272.8	1.7	25.4	86.1	159.8
Nevada	2,565,382	750.6	7.5	42.7	270.2	430.2
Texas	23,904,380	510.6	5.9	35.3	162.2	307.2
Washington	6,468,424	333.1	2.7	40.6	93.6	196.2
Wyoming	522,830	239.3	3.1	30.6	16.1	189.5

Source: Data from "Crime in the United States, 2007: Violent Crime," Federal Bureau of Investigation, 2008.

vehicle thefts by number of registered vehicles, burglaries by number of house-holds, or gun-related crimes by number of registered guns. This method can also be used to calculate victimization rates, arrest rates, and clearance rates.

Percentage Change

Percentage-change statistics allow researchers to compare data over time and across jurisdictions. The computation for this is rather straightforward: sub-tract the earlier number from the later number, and then divide the difference by the earlier number. This statistic allows one to determine whether there has been an increase or decrease in particular crimes. The bottom of Table 12-2 demonstrates this statistic for three index crimes in the United States for three different sets of years.

Trend Analyses

Trend analyses are yet another way of comparing differences over time. One may examine on a histogram how rates have increased for a particular offense, or one may use other means of determining how the data have changed. Trend analyses are quite useful in assessing the impact of crime prevention strategies. The reader should refer to Table 12-2, which shows the number of specific crimes from 1998 to 2007. From this table, one can determine what trends may have existed for these crimes during the reported time frame.

Inferential Statistics

Because of the nature of inferential statistics, to do justice in explaining them requires a text of its own. Further explanations are left for statistics courses and only brief descriptions of selected statistics are offered.

Bivariate Analysis

Bivariate analysis is the examination of the relationship between two vari-ables. Usually this involves attempting to determine how a dependent variable is influenced by an independent variable. The more commonly used bivariate techniques are cross-tabulation (contingency tables) and bivariate (simple linear) regression. In assessing the relationship between the two variables next examined are the contingency table or regression scatterplot results; the measure of association (e.g., gamma or R^2); and the level of statistical signifi-cance (it is hoped 0.05 or lower).

Contingency Tables

With nominal level data, two popular statistical techniques are contingency tables (cross-tabulations) and chi-square, which is a common statistic for a contingency table. A contingency table is a set of interrelated cells. Each cell

TABLE 12-2

Specific Crimes in the United States by Volume, Rate, and Percent Changes, 1998– 2007

Year	Population	Forcible rape volume	Forcible rape rate	Robbery volume	Robbery rate	Aggravated assault volume	Aggravated assault rate
1998	270,296,000	93,144	34.5	447,186	165.4	976,583	361.3
1999	272,691,000	89,411	32.8	409,371	150.1	911,740	334.3
2000	281,421,906	90,178	32	408,016	145	911,706	324
2001	284,796,887	90,491	31.8	422,921	148.5	907,219	318.5
2002	287,973,924	95,235	33.1	420,806	146.1	891,407	309.5
2003	290,788,976	93,883	32.3	414,235	142.5	859,030	295.4
2004	293,656,854	95,089	32.4	401,470	136.7	847,381	288.6
2005	296,507,061	94,347	31.8	417,438	140.8	862,220	290.8
2006	299,398,484	92,757	31	447,403	149.4	860,853	287.5
2007	301,621,157	90,427	30	445,125	147.6	855,856	283.8
Percent change in volume and rate per 100,000 inhabitants for 2, 5, and 10 years							
2007/2006		−2.5	−3.2	−0.5	−1.2	−0.6	−1.3
2007/2003		−3.7	−7.1	+7.5	+3.6	−0.4	−3.9
2007/1998		−2.9	−13	−0.5	−10.8	−12.4	−21.5

can display a variety of data (Table 12-3). From these data, a chi-square statistic is available. Chi-square is one test of statistical significance. These techniques offer a statistical relationship from which one might make an inference.

Bivariate Regression

Bivariate regression, also known as "simple linear regression," is based on the principle that over time things tend to regress toward the mean. For example, if one were to measure the heights of female students enrolled at a school one would find that they range from well above average to well below average height for women. Assuming a normal population, most of the heights would tend to cluster around the mean (average) height. A scatterplot of these heights would most likely enable a line to be drawn showing most heights to be near the mean height.

If one is using interval level data, is dealing with a normal distribution (determined through earlier descriptive statistics), and has a linear relationship, this is an appropriate procedure to see how an independent variable predicts the outcome of a dependent variable. A description of the principles on which simple linear regression is based is not offered here. However, if one has

TABLE 12-3

Example of Contingency/Cross-tabulations: Overall Job Satisfaction by Level of Education/Degree

	Count Exp Val Row Pct Col Pct Tot Pct	EDUCATION				Row Total
		H.S. Diploma 1	Assoc. Deg. 2	Bach. Deg. 3	Master's Deg. 4	
Overall	2	1 .5 100.0% 4.5% 2.3%	0 .2 .0% .0% .0%	0 .3 .0% .0% .0%	0 .0 .0% .0% .0%	1 2.3%
Neutral	3	11 10.2 55.0% 50.0% 25.6%	4 4.2 20.0% 44.4% 9.3%	4 5.1 20.0% 36.4% 9.3%	1 .5 5.0% 100.0% 2.3%	20 46.5%
	4	9 9.2 50.0% 40.9% 20.9%	3 3.8 16.7% 33.3% 7.0%	6 4.6 33.3% 54.5% 14.0%	0 .4 .0% .0% .0%	18 41.9%
Extremely Satisfied	5	1 2.0 25.0% 4.5% 2.3%	2 .8 50.0% 22.2% 4.7%	1 1.0 25.0% 9.1% 2.3%	0 .1 .0% .0% .0%	4 9.3%
Column		22	9	11	1	43
Total		51.2%	20.9%	25.6%	2.3%	100%

Chi-Square	Value	DF	Significance
Pearson	5.12525	9	.82326
Liklihood Ratio	5.53040	9	.78584
Linear-by-Linear Association	.61069	1	.43453

Minimum Expected Frequency—.023

Cells with Expected Frequency < 5 – 13 of 16 (81.3%)

Approximate Statistic	Value	ASE1	Val/ASEO	Significance
Phi	.34524			.82326
Cramer's V	.19933			.82326
Contingency Coefficient	.32634			.82326

a linear relationship, as X (the predictor variable) increases so should Y (the dependent variable). Conversely, if a negative relationship exists (e.g., a crime prevention strategy), as X (the strategy used) increases, Y (the specific crime targeted) it is hoped decreases.

Multivariate Analysis

Multivariate analysis is the examination of the relationship between three or more variables. Usually this involves attempting to determine how a dependent variable is influenced by several independent variables. This methodology offers more insights than bivariate analysis, in that one is able to study the relationships among several variables at one time. The more commonly used multivariate techniques are correlation, multiple regression, Student t test, analysis of variance (ANOVA), discriminate analysis, probit regression, factor analysis, and path analysis (Frankfort-Nachmias & Leon-Guerrero, 2008; Gavin, 2008; Walker & Maddan, 2009; Weisburd & Britt, 2007).

In assessing the relationship between the multiple variables next examined are the correlation table, regression scatterplot results, or other indicators (depending on the technique used); the measure of association (e.g., Pearson r, R^2, Wilks lambda, or other appropriate measure); and the level of statistical significance. A number of multivariate examples are provided in this chapter.

Student t Test

The Student t test is used to compare groups' means for a particular variable and hypothesis testing. Computing Student t is a fairly complex process that contrasts expected outcomes with observed outcomes. The differences among the means of the variables are then assessed. The Student t test provides means for each variable by group and then offers a statistic called the t-value, which indicates whether the relationship between the groups is statistically significant (Table 12-4). As stressed previously, the purpose here is not to teach how to compute this measure, but to inform one of what to look for in the works of others, and what to discuss with a statistician if using this strategy. As seen in Table 12-4 the differences between the means are obvious. The t-values do not mean anything at this point, but indicate the values used by the computer to calculate the significance.

Correlation

Correlation is a commonly used technique for evaluating interval- and ratio-level data. The previous discussion of bivariate regression explained how relationships between two variables are examined based on the assumption of a linear relationship. In correlation, relationships are assessed based on covariation. Covariation simply means that as changes occur in one variable, X, they

TABLE 12-4

Student's *t* for Student's Perceptions of Policing Comparison of Selected Means Scores, by Time

Variable	Mean (*t1*)	Mean (*t2*)	*t*-value
Primary Role	.554	−.458	5.49*
Level of Competency	−.747	−1.289	4.29*
Serve and Protect	−1.000	−.800	−1.27
Corrupt Act	−.598	−.390	−1.33
Strike a Minority	−.171	−.500	2.37*
Ignore Needs	−1.000	−1.060	.55
Preventing Crime	−.476	−.951	2.97*
Harass or Help	−1.374	−1.470	.96
Unknown Reaction	.482	.716	−1.65
Professionalism	.183	.598	−3.38*
Help Society	.627	.928	−1.84
Benefit of Doubt	.256	.646	−2.39*

* = p < .05

Note: *t*1 and *t*2 represent the distribution of the questionnaire. *t*1 was at the beginning of the semester, and *t*2 was the end of the semester.

Source: Modified from Dantzker, M. L. and Ali-Jackson, N. Examining students' perceptions of policing and the effect of completing a policerelated course. In Dantzker, M. L., et al. *Practical Applications for Criminal Justice Statistics*. Butterworth-Heinemann. Copyright Elsevier 1998.

will also change in another variable, Y. A positive correlation reveals that as X values increase, Y values also increase. Assessing correlations is based on how the variation in one value corresponds to variation in the other value.

The means of assessing the correlation of interval-level data is Pearson product-moment correlation. Pearson *r* is used to determine the strength of the association among variables by dividing the covariance of X and Y by the product of the standard deviation of X and Y. The computer calculates these numbers. Of interest here is the direction (recall one is looking at a number between −1 and +1), the strength, and the significance of *r*. Table 12-5 is an example of correlation in analyzing interval data.

Analysis of Variance

ANOVA is another means of examining interval level data. Where correlation uses Pearson *r* to determine the nature of relationships among variables, ANOVA uses something known as an "F ratio" to compare the means of groups. This technique helps avoid committing errors that might occur when using multiple *t* tests (Frankfort-Nachmias & Leon-Guerrero, 2008). ANOVA

TABLE 12-5

Correlation

Variable(s)	Incomp	Motto	Corrupt	College
Role				
Pearson's r	.119	.133	.058	−.145
Sig. (2-tailed)	.013	.006	.230	.003
N	439	435	437	413
Incomp				
Pearson's r	1.000	.346	.308	−.054
Sig. (2-tailed)		.000	.000	.277
N	440	436	438	414
Motto				
Pearson's r	.346	1.000	.349	−.085
Sig. (2-tailed)	.000		.000	.086
N	436	436	434	410
Corrupt				
Pearson's r	.308	.349	1.000	−.106
Sig. (2-tailed)	.000	.000		.032
N	438	434	438	412
College Years				
Pearson's r	−.054	−.085	−.106	1.000
Sig. (2-tailed)	.277	.086	.032	
N	414	410	412	414

allows statisticians to determine significance by assessing the variability of group means. Therefore, this is a useful method for evaluating grouped data (e.g., the outcomes of correctional treatments on inmate groups). Table 12-6 is an example of ANOVA findings.

Multiple Regression

Multiple regression, sometimes referred to by statisticians as "ordinary least squares," is yet another means of evaluating interval-level data. It is based on the same assumptions as bivariate regression. However, instead of assessing the relationship between only two variables, multiple regression usually examines several variables at once. This popular technique enables researchers to look not only at how the independent variables predict the outcome of the dependent variable, but also at the relationships among the

TABLE 12-6

Selected ANOVA Results from a Perceptions Study

Value Label	Mean	Std Dev	Sum of Sq	d.f.	F
AVOID					
Springfield Acad	−1.3750	.5310	13.2500		
MA Regional Acad	−1.7500	.4935	9.5000	1	11.60*
CORRUPT					
Springfield Acad	−1.4375	.7693	27.8125		
MA Regional Acad	−1.4000	.9282	33.6000	1	.04
DRUGS					
Springfield Acad	.0625	1.4790	102.8125		
MA Regional Acad	.4500	1.4313	79.9000	1	1.54
EXPECTATIONS					
Springfield Acad	.1702	1.2036	66.6383		
MA Regional Acad	−.2500	1.0801	45.5000	1	2.89
HELP SOCIETY					
Springfield Acad	1.3750	.7889	29.2500		
MA Regional Acad	1.4000	.8412	27.6000	1	.02
IGNORE					
Springfield	1.3958	.8440	33.4792		
MA Regional Acad	−1.6500	.6222	15.1000	1	2.50

*$p - .01$

independent variables. Multiple regression builds on bivariate regression and correlation by examining partial correlations. It uses variance to assess the ability of an independent variable to predict the dependent variable. Multiple regression goes beyond this analysis by assessing variance among the independent variables (Keith, 2005). The influence on total variation that additional variables cause on the prior variables is known as "partial correlation." One can see how the variables interact, and can use this knowledge to add or remove independent variables from the regression equation. Computers do this in a process known as "step-wise regression."

As in previous sections, one does not have to understand the process to be able to interpret multiple regression. Look for the same criteria as in bivariate regression: the direction, the strength, and the significance of R^2 are important. Additionally, the relationships between the independent variables must be

TABLE 12-7

Multiple Regression: Regressions of County Population Against Crime and Arrests in Montgomery County, 1970–1990

Variable	bo	b1	R^2	F	Sig F
Part I Crimes	1,810.7	38.3	.50	18.7	.000
Part II Crimes	3,029.4	56.4	.29	7.8	.012
Total Crimes	4,840.1	94.7	.50	18.6	.000
Part I Arrests	3,177.5	2.6	.40	12.6	.002
Part II Arrests	−1,059.8	11.9	.47	16.9	.001
Total Arrests	2,117.7	14.6	.60	28.0	.000

Source: Modified from Guynes, R. and McEwen, T. Regression analysis applied to local correctional systems. In Dantzker, M. L., et al. *Practical Applications for Criminal Justice Statistics.* Butterworth-Heinemann. Copyright Elsevier 1998.

examined by reviewing their correlation coefficients. However, to determine whether to use multiple regression and how to deal with problems that may arise, one should consult with a statistician or someone experienced in using regression. An example of multiple regression results is shown in Table 12-7.

Other Multivariate Techniques

In addition to the previous multivariate techniques, there are several others that are popular among criminologists. All of them require a solid understanding of the statistical procedures involved. Probit analysis (also referred to as "probit regression" or just "probit") is similar to, yet different, from multiple regression. It is in effect a cumulative distribution regression model that helps straighten out an S-shaped distribution (Keith, 2005). It is also appropriate when one has interval independent variables but an ordinal-level dependent variable (Keith, 2005).

Logit analysis is also known as "logit" and "logistic regression." Like probit analysis, it is a type of regression that may be used to deal with S-shaped distributions. Logistic analysis is a sophisticated technique used by statisticians when the cumulative distribution is logistic rather than normal as in probit analysis. Like regression, both logit and probit results are assessed using R^2 as the measure of association (Keith, 2005).

Discriminate analysis is a favorite technique of the authors. It is appropriate when one has interval independent variables and a nominal-level dependent variable. Regression is based on the ability of the independent variables to predict the dependent variable. Discriminate analysis focuses on the ability to classify observations to the nominal categories of the dependent variable based on their values on a set of independent variables. The measure of association for discriminate analysis is Wilks lambda.

Factor analysis also categorizes data. It is used to determine patterns among the variation of values of the variables being studied. Variables that are highly correlated are clustered together based on computer-generated factors (Keith, 2005). This is an extremely complex procedure that must later be interpreted by the researcher to determine whether the factor loadings have logical meaning.

Path analysis seeks to provide a graphic depiction of the causal relationships among the independent variables to explain their influences on the dependent variable. Like factor analysis, it is a complex procedure that is best left to statisticians.

Nonparametric Techniques

When the distribution of the data is not normal, standard statistical techniques are usually not appropriate. In those cases, other procedures that are used include chi-square (based on the statistic discussed previously); nonparametric correlation; and nonlinear regression. The strategies used in nonparametric analysis are complex. The reader is advised to consult with a statistician in these circumstances. They are mentioned here so that the reader knows that there are techniques that are available should one have data that are not in a normal distribution.

Summary

This chapter provides an overview of several statistical techniques that should aid the reader in understanding research and preparing their own. The authors do not claim to give one the knowledge needed for in–depth data interpretation (that is provided in statistics texts), but one should be able to grasp the principles involved in conducting inferential statistics. Today's reality is that researchers do not really need to be expert statisticians. Many statistical analyses can be completed through a variety of user-friendly software packages. Often all the researcher needs to do is be able to code and enter data, choose what statistical techniques should be run, interpret the results, and report the findings. To attain a better understanding of statistical analyses without a statistics course, it is suggested that one read almost any criminal justice or criminologic study, focusing particularly on the statistics portion.

METHODOLOGICAL QUERIES

1 The sheriff realizes he really doesn't understand statistics. You try to help him by differentiating for him descriptive, comparative, and inferential statistics.

2 You find you have to explain to him what is meant by statistical significance. Also describe how tests of significance are used.

3 To help him better understand statistics, you explain and provide him with examples from the proposed study of:

   a bivariate analysis,

   b multivariate statistics, and

   c nonparametric techniques.

Writing the Research

What You Should Know!

At this point the reader has been introduced to all the steps for conducting a research study. Collecting the data is only half the battle for many researchers. Writing up the results for presentation or publication can often cause anxiety for the first-time researcher and the seasoned veteran. However, when following the steps offered in this chapter, writing up the results does not have to seem so daunting. After studying this chapter the reader should be able to do the following:

1. Explain the purpose of the title page and describe its structure.
2. Explain the purpose of the abstract and describe its contents.
3. Explain the purpose of the introduction and describe its contents.
4. Explain the purpose of the literature review and describe its contents.
5. Explain the purpose of the methodology section and describe its contents.
6. Explain the purpose of the results section and describe its contents.
7. Explain the purpose of the conclusions section and describe its contents.
8. Explain the purpose of the references and describe their structure.
9. Explain the purpose of tables and figures and describe their contents.
10. Explain the purpose of appendices and describe their contents.

The Research Paper

The topic has been chosen, the design implemented, the questionnaire constructed, the data collected and analyzed, and now one thinks, "Here comes the hard part." For many people, students and scholars alike, conducting the research is seen as the easy or fun part and would be great to do if it did not have to be written up. Experience suggests that students, in particular, really can get into designing and collecting the data, but then they fear the analyses and loathe having to provide a written explanation. However, because the written project is generally the required final goal, whether for fulfilling a course requirement or attempting to get published, it must be taken as seriously as the research itself. Furthermore, the transition from data to words is not as difficult as it may seem (Bachman & Schutt, 2008; Bickman & Rog, 2009; Kline, 2009; Morgan, Reichert, & Harrison, 2002).

Because each person has his or her own writing style, the authors do not try and tell the reader how to go about writing. They do, however, suggest an order for one's paper and what should be included. The basic order should be title page, abstract, introduction, methodology, results, conclusions, references, and appendices. This chapter provides not only an overview of what should be included within each of these components, but also an American Psychological Association (APA) writing style example from a previously unpublished paper. It is important to note that before writing a final paper, one should consult the most current writing manual of the writing style that is to be used (e.g., the APA style manual was in its sixth edition as of the revising of this text).

The Title Page

The title of a research paper, article, thesis, or dissertation should tell the reader in clear and concise terms what the research is about. Often, individuals who are perusing journals or article abstracts do not have (or do not take) the time to read the abstracts. The title is what draws their attention to the paper as possibly being of interest to them. The author's identity and affiliation then follow. If the research has been funded by an external organization, then that is also indicated on the title page. The journal, organization, or instructor to whom the completed research is being submitted may have specific requirements as to how this page is to be structured. An example of a title page can be found in Box 13-1.

Abstract

The abstract is a summary or synopsis of the information being presented in the paper, and starts with the research title. The abstract presents the paper's

---

Box 13-1

Example of a Title Page

Personality Testing of Police Candidates: What's Being Used and Why

M. L. Dantzker, Ph.D.

Dept of Criminal Justice
UTPA
Edinburg, TX

Box 13-2

Example of an Abstract

Despite the current number of states that require psychological screening for police candidates, there is no consistent process. The lack of a consistent or standardized process has been blamed for some police candidates being rejected by one psychologist only to be approved by another psychologist. This presentation offers results from a national study examining what clinical psychologists are using to conduct the pre-employment psychological assessment and their reasons for the use.

---

major arguments and describes the methods used. Note that this abstract is limited to one paragraph. Generally, the abstract is between 100 and 150 words in length. An example of an abstract can be found in Box 13-2.

The Introduction

This section varies in length, depending on whether there is a page limit or on how much information is really available. Regardless of length, this section needs to establish the research problem, the literature that supports its existence, and the reason to research the problem. Furthermore, this section must report the research questions and hypotheses, and usually briefly describes what was done and how.

The introduction and literature review may be contained within one section or broken into two sections. If these parts are separated, the researcher uses the introduction to introduce the reader to the topic and research question. Whether separated or not, the introductory paragraph should alert the reader as to what to expect from the paper. This is followed by the main portion or the literature review, where the writer offers a body of support for the research.

In some cases, there may be very little support; if so, that fact must be acknowledged. In other cases, the writer has to decide how to limit the extensive amount of evidence that is available. Finally, the literature-review section should end with a summary paragraph that indicates what research is being reported and how it was accomplished. An example can be found in Box 13-3.

Box 13-3

Example of an Introduction and Literature Review

According to authoritative sources (Federal Law Enforcement Training Center, 2009; Reeves & Goldberg, 1996) there are more than nineteen thousand police agencies in the United States employing more than one million full time police officers (Crime in the United States, 2008; Sourcebook of Criminal Justice Statistics Online, 2008; State and Local Law Enforcement Statistics, 2008), a sizable part of the entire population of the United States. Moreover, any individual in the United States could potentially interact with a police officer on any given day. Therefore, it could be argued that the psychological characteristics of those selected to become police officers are or should be of importance to every individual in this country. It follows that those who evaluate the psychological characteristics of police officers, along with the process used, should be the best available.

Moreover, being a police officer is a difficult career (Alpert, Dunham, & Stroshine, 2005; Birzer & Roberson, 2006; Koper, 2004; Orrick, 2008). Police officers frequently deal with the more negative aspects of human nature and often must do so with a non-emotional or objective approach, while deeply constrained by law and policy (see e.g., Buetler, Nussbaum, & Meredith, 1988; Dantzker, 2005; Koper, 2004; Orrick, 2008; Varela, Scogin, & Vipperman, 1999). As Buetler et al. (1988) argued the unique nature of policing demands a selection process to find the individuals who are capable of coping with the demands of police work in a satisfactory manner. Hargrove and Hiatt (1989) noted that police officers have constant contact with people and are often in volatile situations; thus, they concluded the assessment of an applicant's interpersonal skills is important in the screening process. Issues of police use of force, corruption, and suicide raise questions about the psychological well-being of those who become police officers (Cochrane, Tett, and Vandecreek, 2003). Finally, Tanigoshi, Kontos, and Remley (2008) suggested law enforcement is one of the most dangerous, stressful, and health-threatening occupations in existence.

It is evident that those selected to become police officers should be thoroughly screened through pre-employment evaluations prior to being hired as police officers or police officer trainees (Corey & Honig, 2008; Miller, 2006; Parisher, Rios, & Reilly, 1979; Reiser, 1973, 1982; Scrivner & Kurke, 1995). However, while psychological testing in screening potential police recruits was first recommended and used on a very limited basis as early as the 1930s, and apparently is popular today (Aumiller, Corey, Allen, et al., 2007; Corey & Honig, 2008; Craig, 2005; Curren, 1998; Dantzker, 2005; Miller, 2006b; Parisher, Rios, & Reilly, 1979; Reiser, 1973, 1982; Scrivner & Kurke, 1995; Zakhary, 2007), use of psychologists for pre-employment police screening first gained the most attention when it was recommended by the National Advisory Commission on Criminal Justice Standards and Goals (1973). Standard 20.1 Entry level Physical and Psychological Examinations states

Box 13-3

Every police agency should require all applicants for police officer positions to undergo thorough entry-level physical and psychological exams to insure detection of conditions that might prevent maximum performance under rigorous physical and mental stress.

I.   Every agency, by 1975, should furnish and require, as a condition of employment, that each applicant pass through a physical and psychological exam. This exam should ...

c.   Include a psychological evaluation conducted under the supervision of a **licensed, competent psychologist or psychiatrist** (bold added by author) (p. 498)

Since then, it appears the number of police agencies requiring psychological pre-employment evaluations increased for many years (Hickman & Reaves, 2006) reaching a total of about 67% of all police agencies by 2004, the latest year for which such information is available (Hickman & Reaves, 2006; IADLEST, 2005). According to the same sources, just over half (26) of all states required statewide pre-employment testing of police recruit candidates.

A study by Cochrane, Tett, and Vandecreek (2003) of 355 police agencies determined that more than 90% of those agencies required a psychological evaluation of applicants. However, they also found that a majority of the 90% were not following public policy or psychological assessment guidelines and recommendations previously proposed by commissions, such as using psychologists or testing for personality (Cochrane et al., 2003). In their 2006 study Hickman and Reaves found that only 67% of local police agencies required pre-employment psychological evaluations and only 26% used a personality inventory as part of the screening. In fact, the majority of agencies using a personality inventory indicated it as a separate category from the psychological evaluations.

It appears that a substantial number of law enforcement agencies are using pre-employment screening in officer hiring processes. This is certainly in keeping with the findings of Macan, Avedon, Paese, and Smith (1995) which identified the attraction of the most qualified applicants to be a key goal of any organization. However, they observed that research on police applicant selection has predominantly focused on psychometric selection. Furthermore, they concluded that the assessments tended to rule applicants out rather than screen them in. That is, the assessment tools used were intended to eliminate clearly unsuitable candidates rather than to identify those likely to perform satisfactorily as law enforcement officers. Consequently, it appears hiring of the most qualified applicants would rely, in part, on systematic and appropriate use of the best practices in psychological practice in the law enforcement environment (Decker, 2006; Rostow & Davis, 2004).

There is no nationally recognized and generally followed set of recommendations as to what questionnaire(s) or evaluative protocols should be used in doing pre-employment screening of law enforcement officers, whether for "screening-out" those who are psychologically unfit for such work or for identifying and "screening-in" those candidates who are likely to perform satisfactorily in the role of police officer (Dantzker, 2005; Dantzker & McCoy, 2006; Decker, 2006; Dempsey & Forst, 2007; Peak, 2008 Rostow & Davis, 2004). Therein lies a problem and a potential means to begin solution of that problem.

Continued

> ### Box 13-3
>
> This paper presents findings from a national study of psychologists who self-identi-fied as providing services to police agencies, to include pre-employment screening, as either an in-house psychologist or full-time consultant. Despite the small sample size, the breadth of the sample suggests a consistency throughout the country supporting the current literature of no agreed upon means of conducting police pre-employment psychological screening.

Methodology

Despite the relevance and importance of the introductory section, one should consider the methodology section as the mainstay of the paper. The writer should discuss the hypothesis or hypotheses, the research design, and the data-gathering technique. This includes explaining the research population, the sampling frame, and the questionnaire or other method used to gather the data. It is within this section that the researcher can fully explain where, when, how, and why the data were attained and analyzed. The methodology section may be written up in one complete section, or it may be subdivided. An example can be found in Box 13-4.

> ### Box 13-4
>
> **Example of a Methodology Section**
>
> While there are a multitude of methods available to conduct research, one of the most popular social science approaches is through survey research (Banyard & Grayson, 2009; Dantzker & Hunter, 2006; Gillham, 2009; Gravetter& Forzano, 2009). This study used the survey method with a modern twist, use of the Internet. The questionnaire used was created specifically for this research focus and was made available to poten-tial respondents through SurveyMonkey.Com, which has been used by the American Psychological Association itself to conduct survey research. The data was analyzed using descriptive and comparative statistical methods.
>
> The data for this study was collected through a questionnaire created specifically for this research. It consisted of thirty-three items accessible through the Internet site of SurveyMonkey.com. The items were related to the independent and dependent variables.

Results

Whether this is a separate section or part of the methodology section is actu-ally a matter of preference. Either way, in this section the writer describes the sample's characteristics, the statistical techniques used, the results, and whether they supported the hypotheses. It is in this section that various tables, graphs, and charts are commonly used to describe the data. An example can be found in Box 13-5.

Box 13-5

Example of a Results Section

The target sample was a combination of individuals randomly drawn from members of the APA belonging to Division 12 (n=1000), and all members at the time of the Police and Public Health section of Division 18 (n=204) and all U.S. members of the Police Psychology section of the IACP (n=188). The total number of individuals contacted via e-mail with a request to participate in the study was 1392.

From among the 1392, 39 individuals responded by e-mail stating that they would not participate in the study and 94 e-mails were returned as undeliverable. Subtracting the non-participants and undeliverable e-mails produced a respondent sample of 1259. Of the 1259, 228 (18%) individuals responded to the survey with 181 (79%) fully completed questionnaires. Among the 181 completed surveys, 77 (43%) met the specific criteria sought for comparison with 40 (52%) being identified as Clinical Psychologists and 37 (48%) falling under the Police Psychologist label. Thirty states were represented with two respondents indicating they provided their services nationally.

To maintain anonymity a minimum number of descriptors were obtained. The four of main interest in this study were gender, organizational membership, years providing services to police agencies, and the type of agency served. Males respondents were dominant (80%), 57% belonged to both APA and IACP, the majority (60%) had been providing these services for over 10 years, and the majority (63%) provided these services to more than one type of agency. (See Table One)

A primary question of interest regarding the pre-screening psychological process is whether it should be for screening out candidates or screening them in. In response to ranking the reasons for the evaluation 71% ranked to screen out as the number one reason while only 21% ranked to screen in as the number one reason. (See Table Two)

The main question of this study was what evaluative measures the respondents used and the reason for its use. Based on the literature four major categories of measures were represented: intelligence, aptitude, personality, and projectives. For each category an "other" response was also available. Twelve specific measures, divided among the four categories, were provided. Among the twelve measures, only five were identified as being used, four of which were personality measures. The other specific measure was an intelligence test. As for the reasons for their use, respondents had seven choices: research support, predictability, mandated, cost, job validity, norming, and other. The choices were collapsed into two categories: research support and other.

Beginning with intelligence testing, 13 respondents indicated they use the WAISIII/IV, 12 of whom do so because of the research support. Twenty-six respondents indicated using some type of IQ test, 21 because of the research support. For aptitude, only four respondents indicated using any type of measure, but no reason was given. (See Table Three)

Personality testing has the longest history within police psychological screening. The MMPI has been the most written about and most often cited for its use in police screening. Results from this study continue to support its popularity. Among the four specific personality measures, the MMPI was indicated as being used by 50 respondents (over 60%) with 46 choosing research support as their reason for its use. The

Continued

Box 13-5

next most used measure was the CPI. Despite that the four measures provided tended to be among the most often cited in research 26 respondents indicated using a personality test not listed such as the Inwald Personality Inventory which is one of the few normed specifically for law enforcement/policing. Finally 25 respondents indicated using a short sentence completion test because of research support while five use a projective measure. (See Table Three)

The last question of interest was whether respondents would support the creation of a standardized approach to conducting pre-employment screening. Respondents were given five choices: Would Support it, Great Idea, Reasonable Idea, Mixed Feelings, and Terrible Idea. Interestingly only 17 (23%) thought it was a terrible idea. (See Table Four)

Conclusions

Usually the last section of the research paper is the conclusion. Generally, this section is used to offer insights about the research, and whether it did what was expected, along with any possible problems. This section can also be used to discuss implications of the research and to provide a forum for suggestions for future related research. Some authors choose to divide this information into two sections: discussion and conclusions. There is no wrong or right here, it is simply a matter of preference. An example follows:

Box 13-6

Example of a Conclusion

Despite the amount of research published related to police candidate pre-employment screening, there still seems to lack research with a strong direction toward consensus and/or standardization of the process. What should be used and why remains a hot topic for debate. This study's results tended to support previous research showing an array of items used but primarily because of the research support which doesn't necessarily prove the item is the best measure to use for police pre-employment screening. Furthermore, there is the question of whether this process should be one of screening out or screening in candidates. The measures identified in this study are primarily used for screening out which leads to the question, would these same measures adequately work to screen in candidates? Obviously, more research is required perhaps beginning with answering the question, to screen in or screen out. Answering this question would certainly assist in addressing what measures should be used for the process and whether standardizing the process is even possible.

References or Bibliography

Unless the topic has never been addressed in research before (a very unlikely event), there are always some sources that help to support the research. These sources are what help establish the literature review. When using other sources (whether quoting, paraphrasing, or simply as an outlet for affirming what is already known) the source should be recognized. Not only can this be done throughout the paper by various citation methods, but it must be done at the end in the form of a reference list or bibliography. The format depends on the source to which the paper is being submitted (i.e., publication, instructor, organization). Failing to cite sources can lead to charges of plagiarism, something every writer should strive to avoid.

Most papers written for criminal justice and criminology follow the APA style for reference citing in the text and at the end of the paper. Reviewing the introduction section, one observes how two different formats are used. One is for paraphrasing or using someone else's ideas and giving credit (name[s] of the author[s], publication date) and the other is for direct quotes (name[s] of the author[s], publication date, page number[s]). For the reference list, only those sources cited in the body of the text should be included. The information listed should include the name(s) of the author(s); the year of publication; the name of the publication (if it is an article, the article title comes first, followed by the publication title); if it is a book, the city, state, and name of publisher; or if a journal, the volume and issue number and the page number(s). An example can be found in Box 13-7.

Tables and Figures

Most likely, the research paper includes at least one table or figure. As previously noted, these items can be placed within the text, kept separate and noted where they should be inserted, or else placed at the end of the paper. Regardless of their location, tables and figures need to be clear as to their content and readily understandable. Tables 13-1, 13-2, 13-3, and 13-4 are from the example paper illustrated throughout this chapter.

Appendixes

This last section is not a requirement of every paper. However, it is often useful to include a copy of the questionnaire or other tools that should be shared but do not belong in the body of the paper. There are no limitations to the number of appendixes a paper can have, except for those established by the instructor or the journal.

Box 13-7

Example of a Reference List

Alpert, G. P., Dunham, R. G., & Stroshine, M. S. (2005). *Policing: Continuity and change*. Prospect Heights, IL: Waveland Press.

Aumiller, G. S., Corey, D., Allen, S., Brewster, J., Cuttler, M., Gupton, H., Honig, A. (2007). Defining the field of Police Psychology: Core domains & proficiencies. *Journal of Police and Criminal Psychology, 22*(2), 65–76.

Banyard, P., & Grayson, A. (2009). *Introducing Psychological Research* (3rd ed.). NY: Palgrave Macmillan.

Beutler, L. E., Nussbaum, P. D., & Meredith, K. E. (1988). Changing Personality Patterns of Police Officers. *Professional Psychology: Research and Practice, 19*(5), 503–507.

Birzer, M., & Roberson, C. (2006). *Policing Today and Tomorrow*. Upper Saddle River, NJ: Prentice Hall.

Cochrane, R., Tett, R. P., & Vandecreek, L. (2003). Psychological testing and the selection of police officers: A national survey. *Criminal Justice and Behavior, 30*(5), 511–537.

Corey, D. M., & Honig, A. L. (2008). Police psychology in the 21st century. *The Police Chief, 75*(10), 138.

Craig, R. J. (2005). Police Psychology. Personality-guided forensic psychology, (pp. 55–80). Washington, DC, US: American Psychological Association.

Crime in the United States 2008. (http://www.fbi.gov/ucr/cius2008/data/table_70.html).

Curran, S. F. (1998). Pre-employment psychological evaluation of law enforcement applicants. *The Police Chief, 65*(10), 88–93.

Curran, S. F., & Saxe-Clifford, S. (2004). *Psychological evaluation of public safety applicants: The 2004 Revised Guidelines of the Police Psychological Services Section*. (Available through www.IACP.org).

Dantzker, M. L. (2005). *Understanding Today's Police* (4th Ed.). Monsey, NJ: Criminal Justice Press.

Dantzker, M. L. & Hunter, R. D. (2006). *Research Methods for Criminology and Criminal Justice: A Primer* (2nd ed.). Sudbury, MA: Jones and Bartlett Publishers.

Dantzker, M. L., & McCoy, J. H. (2006). Psychological screening of police recruits: A Texas perspective. *Journal of Police and Criminal Psychology, 21*(1), 23–32.

Decker, K. P. (2006). Introduction to fitness for duty evaluations in law enforcement personnel. In K. P. Decker (Ed.), *Fit, Unfit or Misfit: How to perform fitness for duty evaluations in law enforcement professionals* (pp. 3–31). Springfield, IL: Charles C. Thomas, Publisher, Ltd.

Dempsey, J. S., & Forst, L. S. (2007). *An Introduction to Policing* (4th ed.). Belmont, CA: Wadsworth Publishing.

Edenborough, R. (2005). *Assessment Methods in Recruitment, Selection & Performance*. Sterling, VA: Kogan Page.

Gillham, B. (2009). *Developing a Questionnaire (Real World Research)* (2nd ed.). NY: Continuum International Publishing Group.

Gravetter, F. J., & Forzano, L. B. (2009). *Research Methods for the Behavioral Sciences* (3rd ed.). Belmont, CA: Wadsworth Publishing.

Continued

Box 13-7

Hargrave, G. E., & Hiatt, D. (1989). Use of the California Psychological Inventory in Law Enforcement Officer Selection. *Journal of Personality Assessment, 53*(2), 267–277.

Hickman, M. J., & Reaves, B. A. (2006). *Local Police Departments, 2003.* Washington, D.C.: Bureau of Justice Statistics.

IADLEST (2005). *Sourcebook 2005.* (CD-Rom).

Koper, C. S. (2004). Hiring and Keeping Police Officers. Washington, D. C.: The Office of Justice Programs, National Institute of Justice.

Macan, T. H., Avedon, M. J., Paese, M., & Smith, D. E. (1995). The effects of applicants' reactions to cognitive ability tests and an assessment center. *Personnel Psychology, 47,* 715–739.

Miller, L. (2006). Police Psychology: What is it? *The Doe Report* available at http://www.doereport.com/police_psychology.php

Orrick, W. D. (2008). *Recruitment, Retention, and Turnover of Police Personnel: Reliable, practical, and effective solutions.* Springfield, IL: Charles C. Thomas Publishers.

Parisher, D., Rios, B., & Reilly, R. R. (1979). Psychologists and psychological services in urban police departments—A national survey. *Professional Psychology, 10*(1), 6–7.

Peak, K. J. (2008). *Policing America* (6th ed). Upper Saddle River, NJ: Prentice Hall.

Reiser, M. (1973). The Police Psychologist: A New Role. *Professional Psychology, 4*(2), 119–120.

Rostow, C. D., & Davis, R. D. (2004). A handbook for Pychological Fitness-for-Duty Evaluations in Law Enforcement. New York: The Haworth Clinical Practice Press.

Scrivner, E. M., & Kurke, M. I. (1995). Police psychology at the dawn of the 21st century. In M. I. Kurke and E. M. Scrivner (Eds.), *Police Psychology into the 21st Century,* pp. 3–30. Hillsdale, NJ: Lawrence Erlbaum Associates, Inc.

Sourcebook of Criminal Justice Statistics Online (2008). Available at www.albany.edu/sourcebook

State and Local Law Enforcement Statistics (2008). Available at www.ojp.usdoj.gov/bjs/.

Tanigoshi, H., Kontos, A. P., & Remley, Jr., T. P. (2008). The effectiveness of individual wellness counseling on the wellness of law enforcement officers. *Journal of Counseling & Development, 86*(1), 64–74.

Varela, J. G., Scogin, F. R., & Vipperman, R. K. (1999). Development and preliminary validation of a semi structured interview for the screening of Law Enforcement candidates. *Behavioral Sciences and the Law, 17,* 467–481.

Zakhary, Y. (2007). Police Psychological Services Section. *The Police Chief, 74*(8), 86.

TABLE 13-1

Sample Demographics

Variable	Frequency	Valid Percent
Gender		
Female	22	29.3
Male	53	79.7
Membership		
APA	27	35.5
IACP	06	07.9
Both	43	56.6
Years Providing Psych Services		
< 1	03	03.9
1–5 yrs	15	19.7
6–10 yrs	13	17.1
> 10 yrs	45	59.2
Types of Agencies Served		
Municipal	17	22.7
County	07	22.7
State	04	05.3
Combo	47	62.7

TABLE 13-2

Ranking of Reasons for Preemployment Psych Evaluation

Order	To Screen Out	To Screen In	Legal	Baseline
(1)	53 (70.7)	15 (20.8)	12 (17.1)	01 (01.5)
(2)	18 (24.0)	26 (36.1)	20 (28.6)	04 (06.1)
(3)	03 (04.0)	22 (30.6)	25 (35.7)	12 (18.2)
(4)	01 (0.13)	09 (12.5)	13 (18.6)	48 (72.7)

TABLE 13-3

Measure Used and Reason

Measure	No. that Use	Reason	
		Research	*Other*[1]
WAISIII/IV	13	12	01
OIT*	26	21	05
OAT**	04	n/r	n/r
MMPI	50	46	04
PAI	31	23	08
16PF	25	18	07
CPI	32	23	09
OPT***	26	17	08
SSCT****	25	20	05
Projective*****	05	n/r	n/r

[1]Includes predictability, mandated, cost, validity, norming, and other
*Other Intelligence tests such as the K-Bit, Shipley, Wonderlic, other
**Other Aptitude tests such as ESI, Wonderlic, other
***Other Personality tests such as Hilson, COPS, Inwald, other
****Short Sentence Completion such as Rotter, Sacks, other
*****Rorschach, Edwards, Bender-Gestalt

TABLE 13-4

Standardization of Evaluation Process

Response	Frequency	Valid Percent
Would Support It	20	26.7
Great Idea	01	01.3
Reasonable Idea	04	05.1
Mixed Feelings	33	44.0
Terrible Idea	17	22.7

Summary

Regardless of the fear or loathing one feels about writing up the research, this task is important. It may also be required. Formatting the paper in the manner suggested can make the process easier. Grammar and spelling are extremely important, along with using language that the intended audience can understand. Today, with the availability of various word-processing software, which often includes spelling and grammar checkers, this aspect of the writing should not be as difficult as in years past. Ultimately, the goal should be to submit the most efficiently written paper possible.

The objective of this chapter is to demonstrate the research process through the act of writing up the research into a complete paper, divided into sections to explain each part better. Consequently, one should now have a much better understanding of research methods. It is hoped that the reader finds by this point that conducting and reporting research is not as intimidating as it perhaps was perceived to be at the beginning of this text.

Summing Up

What You Should Know!

This chapter provides an overview of the key elements of research discussed in the previous chapters of this text. The intent is to help the reader see how all of the information provided earlier enables one to progress from the development of a possible research topic to the completion of a methodologically sound research paper. This should refresh one's memory in preparation for a possible final examination in this course. In addition, it aids as a reference for dealing with future research assignments. After studying this chapter the reader should be able to do the following:

1. Present and discuss the issues involved in research: what, why, and how.
2. Recognize and explain the issues involved in research ethics.
3. Present and discuss the issues involved in getting started.
4. Identify and describe the issues involved in the language of research.
5. Present and discuss the issues involved in qualitative research.
6. Recognize and explain the issues involved in quantitative research.
7. Present and discuss the issues involved in research designs.
8. Identify and describe the issues involved in questionnaire construction.

9. Present and discuss the issues involved in sampling.
10. Explain the issues involved in data collection.
11. Recognize and explain the issues involved in data processing and analysis.
12. Present and discuss the issues involved in inferential statistics.
13. Present and discuss the issues involved in writing the research.

Doing Criminologic Research

Ordinary human inquiry may be flawed because of inaccurate observation, overgeneralization, selective observation, and illogical reasoning. The scientific method seeks to prevent the errors of casual inquiry by using procedures that specify objectivity, logic, theoretical understanding, and knowledge of prior research in the development and use of a precise measurement instrument designed to accurately record observations.

Types of Research

Research creates questions, but ultimately, regardless of the subject or topic under study, it is the goal of research to provide answers. Research may occur in one of four formats or types: (1) descriptive, (2) explanatory, (3) predictive, and (4) intervening knowledge. Knowledge that is descriptive allows one to understand what something is. Explanatory research tries to tell why something occurs, the causes behind it. Predictive research gives some foresight as to what may happen if something is implemented or tried. Intervening knowledge allows one to intercede before a problem or issue gets too difficult to address.

Steps in Research

Whether the research is applied or basic, qualitative or quantitative, the basic steps are applicable to each. There are five primary steps in conducting research:

1. Identifying the problem: Identifying or determining the problem, issue, or policy to be studied is what sets the groundwork for the rest of the research.
2. Research design: The research design is the blueprint, which outlines how the research is conducted.
3. Data collection: Regardless of the research design, data collection is a key component. A variety of methods exist. They include surveys, interviews, observations, and previously existing data.
4. Data analysis: Proper analysis and interpretation of the data is integral to the research process.

5. Reporting: The last phase of any research project is the reporting of the findings. Regardless of the audience or the medium used, the findings must be coherent and understandable.

The Language of Research

Research has a language all its own. As with any language, one must know the terminology to understand the language better.

Theory

All criminal-justice practice is grounded in criminologic theory. A theory is a statement that attempts to make sense of reality. Proving that a theory is valid is a common goal of criminologic and criminal justice researchers.

Conceptualization

Concepts are viewed as the beginning point for all research endeavors and are often very broad in nature. They are the bases of theories and serve as a means to communicate, introduce, classify, and build thoughts and ideas. To conduct research, the concept must first be taken from its conceptual or theoretical level to an observational level. This process is known as "conceptualization."

Operationalization

Operationalization is describing how a concept is measured. This process is best described as the conversion of the abstract idea or notion into a measurable item. The primary focus of the operationalization process is the creation of variables and subsequently developing a measurement instrument to assess those variables. Variables are concepts that may be divided into two or more categories or groupings known as "attributes."

Variables

A dependent variable is a factor that requires other factors to cause or influence change. The dependent variable is the outcome factor or that which is being predicted. The independent variable is the influential or the predictor factor. These are the variables believed to cause the change or outcome of the dependent variable and are something the researcher can control.

Hypotheses

Once the concept has been operationalized into variables fitting the theory in question, most research focuses on testing the validity of statements called hypotheses. The hypothesis is a specific statement describing the expected relationship between the independent and dependent variables. There are three common types of hypotheses: (1) research, (2) null, and (3) rival.

Sampling

A sample is a group chosen from within a target population to provide information sought. There are a number of sampling strategies that are available in criminologic research. These include random sampling, stratified random sampling, cluster sampling, snowball sampling, and purposive sampling.

Validity

Validity is a term describing whether the measure used accurately represents the concept it is meant to measure. There are four types of validity, explained below:

1. Face validity: This is the simplest form of validity and refers to whether the measuring device seems, on its face, to measure what the researcher wants to measure. This is primarily a judgmental decision.
2. Content validity: Each item of the measuring device is examined to determine whether each element is measuring the concept in question.
3. Construct validity: This validity inquires as to whether the measuring device does indeed measure what it has been designed to measure. It refers to the fit between theoretical and operational definitions of the concept.
4. Criterion (or pragmatic) validity: This type of validity represents the degree to which the measure relates to external criterion. It can either be concurrent (does the measure enhance the ability to assess the current characteristics of the concept under study) or predictive (the ability to accurately foretell future events or conditions).

Reliability

Reliability refers to how consistent the measuring device would be over time. In other words, if the study is replicated the measuring device provides consistent results. The two key components of reliability are stability and consistency. Stability means the ability to retain accuracy and resist change. Consistency is the ability to yield similar results when replicated.

Data

Data are simply pieces of information gathered from the sample that describe events, beliefs, characteristics, people, or other phenomena. Data may exist at one of four levels, listed below:

1. Nominal data: This level data are categoric based on some defined characteristic. The categories are exclusive and have no logical order. For example, gender is a nominal level data form.
2. Ordinal data: Ordinal data are categorical, but whose characteristics may be rank-ordered. These data categories are also exclusive but are

scaled in a manner representative of the amount of characteristic in question, along some dimension. For example, types of prisons composed as minimum, medium, and maximum.

3. Interval data: Categorical data for which there is a distinctive, yet equal, difference among the characteristics measured are interval data. The categories have order and represent equal units on a scale with no set zero starting point (e.g., the IQ of prisoners).

4. Ratio data: This type of data is ordered; has equal units of distance; and a true zero starting point (e.g., age, weight, income).

Getting Started

As with so many issues in life, often the hardest aspect of research is getting started. There are a number of issues involved in beginning criminologic research. Proper preparation is vital to the successful completion of a research project.

Picking a Topic

Before the topic is chosen, one should consider what currently exists in the literature, any gaps in theory or the current state of art, the feasibility of conducting the research, whether there are any policy implications, and possible funding availability.

Reviewing the Literature

Ultimately, the best way to begin a research effort is to focus on a particular issue, phenomenon, or problem that most interests the individual. In doing so, one must be sure to determine what the problem, issue, or phenomenon is, and organize what is known about it. Once a literature search is completed, the research questions can be formulated.

The Research Question

A well-worded research question should give a clear indication of the outcomes one might expect at the conclusion of the research. After establishing the research questions, the researcher must next offer what specifically is going to be studied and the expected results. This is usually accomplished through statements or propositions referred to as "hypotheses."

Hypotheses

A hypothesis is a specific statement describing the expected result or relationship between the independent and dependent variables. The three most common types are (1) the research hypothesis, which is a statement of the expected relationship between variables offered in a positive manner; (2) the null hypothesis, which is a statement that the relationship or difference being tested does not exist; and (3) the rival hypothesis, which is a statement offering an alternate explanation for the research findings.

Variables

Variables are factors that can change or influence change. They result from the operationalization of a concept. The two types of variables are the dependent and independent. The dependent variable is the factor being influenced to change over which the researcher has no controls. It is the outcome item or what is being predicted. The independent variable is the factor that influences or predicts the outcome of the dependent variable. This variable is something the researcher can control.

Research Ethics

Ethics were defined as doing what is morally and legally right in the conducting of research. This requires the researcher to be knowledgeable about what is being done; use reasoning when making decisions; be both intellectual and truthful in approach and reporting; and consider the consequences, in particular, to be sure that the outcome of the research outweighs any negatives that might occur.

Ethical neutrality requires that the researchers' moral or ethical beliefs are not allowed to influence the gathering of data or the conclusions that are made from analyzing the data. Objectivity is striving to prevent personal ideology or prejudices from influencing the process. In addition to these concerns, the researcher must also ensure that their research concerns do not negatively impact on the safety of others.

Ethical Concerns

Ethical concerns include the following:

1. Harm to others: Physical harm most often can occur during experimental or applied types of research. Psychologic harm might result through the type of information being gathered. Social harm may be inflicted if certain information gathered is released that should not have been.
2. Privacy concerns: Individuals in America have a basic right to privacy. In many cases, research efforts may violate that right. Ethically speaking, if a person does not want his or her life examined, then that right should be granted.
3. Informed consent: Normally, this requires having the individual sign an informed consent form or for the instructions to indicate that the survey

is completely anonymous, voluntary, and that the information is only being used for the purpose of research.

4. Voluntary participation: Participation should be voluntary. If not, there must be valid reasons that can be given showing that the knowledge could not otherwise be reasonably obtained and that no harm will come to the participants from their compulsory involvement. To ensure that informed consent is provided, and to judge the value and ethical nature of the research, many universities have IRBs.

5. Deception: Some types of research (particularly field research that requires the researcher to in essence "go undercover" to gain the knowledge that he or she is seeking) cannot be conducted if the subjects are aware that they are being studied. Such research is controversial and must be carefully thought out before it is undertaken.

Ethical Research Criteria

The following criteria should be followed to produce ethical research: avoid harmful research; be objective in designing, conducting, and evaluating the research; use integrity in the performance and reporting of the research; and protect confidentiality.

Qualitative Research

Qualitative research is defined as the nonnumerical examination and interpretation of observations for the purpose of discovering underlying meanings and patterns of relationships. Such analysis enables researchers to verbalize insights that quantifying of data would not permit. It also allows one to avoid the trap of "false precision," which frequently occurs when subjective numerical assignments are made.

Types of Qualitative Research

Field Interviewing

Field interviewing consists of structured interviews, semistructured interviews, unstructured interviews, and focus groups. A structured interview entails the asking of preestablished open-ended questions of every respondent. A semi-structured interview goes beyond the responses given to the actual questions for a broader understanding of the answer. Unstructured interviews seldom keep to a schedule, and there usually are not any predetermined possible answers. Focus groups interview several individuals in one setting.

Field Observation or Participant–Observer

Field observation consists of observing individuals in their natural setting. The method of observation is determined by the role of the researcher. The full participant method allows the researcher to carry out observational research, but does so in a "covert" manner. The participant researcher participates in the activities of the research environment but is known to the research subjects to be a researcher. The complete researcher avoids all possible interaction with the research subjects.

Ethnographic Study

Ethnographic study or field research overlaps with field observation in that the researcher actually enters the environment under study, but does not necessarily take part in any activities.

Sociometry

Sociometry is a technique by which the researcher can measure social dynamics or relational structures, such as who-likes-whom. Information can be gathered through interviews or by observation and indicates who is chosen and the characteristics about those who do the choosing.

Historiography

Historiography (also known as "historical" or "comparative" research) is the study of actions, events, phenomena, and so forth that have already occurred. This type of research often involves the study of documents and records with information about the topic under study. Historiography can be either qualitative or quantitative in nature depending on the materials being used and the focus of the research.

Content Analysis

Like historical research, content analysis is the study of social artifacts to gain insights about an event or phenomenon. It differs in that the focus is on the coverage of the event by the particular medium being evaluated rather than the event. Depending on how the research is conducted, this may be either qualitative or quantitative in nature.

Quantitative Research

Quantitative research was defined as the numerical representation and manipulation of observations for the purpose of describing and explaining the phenomena that those observations represent. It is based on empiricism. The use of the scientific method with its focus on causation rather than casual observation is what makes empiricism important. It is this emphasis on empiricism rather than idealism that is the basis on which positive criminology is founded.

Causality

In applying empirical observation to criminal justice research the focus is on causal relationships. When one tries to examine numerous explanations for why an event occurred, this is known as "idiographic explanation." When researchers focus on a relatively few observations to provide a partial explanation for an event this is known as "nomothetic causation." Nomothetic explanations of causality are based on probabilities. It is this use of probability that enable one to make inferences based on a relatively few observations.

The Criteria for Causality

The first criterion is that the independent variable (the variable that is providing the influence) must occur before the dependent variable (the variable that is being acted on). The second criterion is that a relationship between the independent and dependent variable must be observed. The third criterion is that the apparent relationship is not explained by a third variable.

Necessary and Sufficient Cause

In investigating causality one must meet the previously mentioned criteria but does not have to demonstrate a perfect correlation. If a condition or event must occur for another event to take place, that is known as a "necessary cause." The cause must be present for the effect to occur. When the presence of a condition ordinarily causes the effect to occur, this is known as a "sufficient cause." The cause usually, but not always, creates the effect.

False Precision

When quantifying data it is imperative that the numerical assignments be valid. If one arbitrarily assigns numbers to variables without a logical reason for doing so, the numbers have no true meaning. This assignment is known as "false precision." One has quantified a concept but that assignment is subjective rather than objective. The precision that is claimed really does not exist.

Types of Quantitative Research

Survey Research

One of the most popular research methods in criminal justice is the survey. The survey design is used when researchers are interested in the experiences, attitudes, perceptions, or beliefs of individuals, or when trying to determine the extent of a policy, procedure, or action among a specific group. Surveys may consist of personal interviews, mail questionnaires, or telephone surveys.

Field Research

If field observations were made (e.g., observing how many vehicles ran a certain stop sign in a given time period or how many times members of the group being observed exhibited specific behaviors) that allowed for numerical assignments, then this would be quantified field research.

Unobtrusive Research

Unobtrusive research is research that does not disturb or intrude into the lives of human subjects to obtain research information. Examples of quantitative unobtrusive research include analysis of existing data, historical research, and content analysis. In the analysis of existing data the researcher obtains the existing data and reanalyzes that data. Historical research and content analysis are conducted as discussed previously, but emphasize statistical rather than verbal analysis.

Evaluation Research

The evaluation research is a quantified comparative research design that assists in the development of new skills or approaches. It aids in the solving of problems with direct implications to the "real world." This type of research usually has a quasiexperimental perspective. Evaluation research that studies existing programs is frequently referred to as "program evaluation."

Combination Research

Combinations of different quantitative methods are commonly used. They may include the use of survey research and field observation, survey research and unobtrusive research, unobtrusive research and field observation, or a combination of all three.

Research Designs

There are a number of issues to consider in selecting a research design. These include the following:

1. Purpose of research: The purpose of the research project. It should be clearly indicative of what will be studied.
2. Prior research: Review similar or relevant research. This promotes knowledge of the literature.
3. Theoretical orientation: Describe the theoretical framework on which the research is based.
4. Concept definition: List the various concepts that have been developed and clarify their meanings.
5. Research hypotheses: Develop the various hypotheses that will be evaluated in the research.
6. Unit of analysis: Describe the particular objects, individuals, or entities that are being studied as elements of the population.
7. Data collection: Determine how the data are to be collected. Who will collect it, who will be studied, and how will it be done?

8. Sampling procedures: Sample type, sample size, and the specific procedures to be used.

9. Instruments used: The nature of the measurement instrument or data collection device that is used.

10. Analytic techniques: How the data will be processed and examined. What specific statistical procedures will be used.

11. Time frame: The period of time covered by the study. This includes the time period examined by research questions and the amount of time spent in preparation, data collection, data analysis, and presentation.

12. Ethical issues: Addresses any concerns as to the potential harm that could occur to participants. Also deals with any potential biases or conflicts of interest that could impact on the study.

Types of Design

There are a number of designs used by criminologic researchers.

1. Historical designs allow the researcher systematically and objectively to reconstruct the past. This is accomplished through the collection, evaluation, verification, and synthesis of information, usually secondary data already existing in previously gathered records, to establish facts.

2. Descriptive research design focuses on the describing of facts and characteristics of a given population, issue, policy, or any given area of interest in a systematic and accurate manner. Like the historical design, a descriptive study can also rely on secondary or records data.

3. Developmental or time series research design allows for the investigation of specifically identified patterns and events, growth, or change over a specific amount of time. There are several time-series designs that are available: cross-sectional studies, longitudinal studies, trend studies, cohort studies, and panel studies.

4. The primary concept of the cross-sectional design is that it allows for a complete description of a single entity during a specific time frame.

5. Where cross-sectional studies view events or phenomena at one time, longitudinal studies examine events over an extended period. For this reason, longitudinal studies are useful for explanation and exploration and description.

6. Trend studies examine changes in a general population over time. For example, one might compare results from several census studies to determine what demographic changes have occurred within that population.

7. Cohort studies are trend studies that focus on the changes that occur in specific subpopulations over time. Usually cohort studies use age groupings.

8. Panel studies are similar to trend and cohort studies except that they study the same set of people each time. By using the same individuals, couples, groups, and so forth, researchers are able more precisely to examine the extent of changes and the events that influenced them.

9. The case (sometimes referred to as "case and field") research design allows for the intensive study of a given issue, policy, or group in its social context at one point in time even though that period may span months or years. It includes close scrutiny of the background, current status, and relationships or interactions of the topic under study.

10. Correlational, one of the more popular research designs, allows researchers to investigate how one factor may affect or influence another factor, or how the one factor correlates with another. In particular, this type of design focuses on how variations of one variable correspond with variations of other variables.

11. Causal–comparative design allows the researcher to examine relationships from a cause-and-effect perspective. This is done through the observation of an existing outcome or consequence and searching back through the data for plausible causal factors.

12. True or classic experimental design allows for the investigation of possible cause-and-effect relationships where one or more experimental units will be exposed to one or more treatment conditions. The outcomes are then compared to the outcomes of one or more control groups that did not receive the treatment. This design includes three major components: (1) independent and dependent variables, (2) experimental and control groups, and (3) pretesting and posttesting.

13. Quasiexperimental design is unlike the true experimental design where the researcher has almost complete control over relevant variables. The quasiexperimental design allows for the approximation of conditions similar to the true experiment. However, the setting does not allow for control or manipulation of the relevant variables.

Questionnaire Construction

Whenever possible use a questionnaire that has previously been developed and tested. The primary reason for this is because it eliminates the worries of validity and reliability, two major concerns of questionnaire development. In creating a new survey instrument there are several aspects for consideration including reliability and validity and the level of measurement used.

1. Start with a list of all the items one is interested in knowing about the group, concept, or phenomenon.

2. Be prepared to establish validity and reliability.

3. The wording in the questionnaire must be appropriate for the target audience.

4. Be sure that it is clearly identifiable as to who should answer the questions.

5. Avoid asking any questions that are biased, leading, threatening, or double-barreled in nature.

6. Before construction, a decision must be made whether to use open- or closed-ended questions, or a combination.

7. Consider that the respondents may not have all the general information needed to complete the questionnaire.

8. Whenever possible pretest the questionnaire before it is officially used.

9. Set up questions so that the responses are easily recognizable whether it is self-administered or interview.

10. The questionnaire should be organized in a concise manner that keeps the interest of the respondent, encouraging him or her to complete the entire questionnaire.

Scales

A common element of survey research is the construction of scales. A scale can either be the measurement device for a question or statement or a compilation of statements or questions used to represent or acknowledge change. The items making up the scale need to represent one dimension befitting a continuum that is supposed to be reflective of only one concept. The representativeness of any scale relies greatly on the level of measurement used.

Arbitrary scales are designed to measure what the researcher believes it is measuring and is based on face validity (discussed previously) and professional judgment. Although this allows for the creation of many different scales, it is easily criticized for its lack of substantive support.

Attitudinal scales are more commonly found in criminal justice and criminologic research. There are three primary types available: (1) Thurstone, (2) Likert, and (3) Guttman. The construction of a Thurstone scale relies on the use of others (sometimes referred to as "judges") to indicate what items they think best fit the concept. There are two methods for completing this task. The first method is the "paired comparisons." Here the judges are provided several pairs of questions or statements and asked to choose which most favorably fits the concept under study. The questions or statements picked most often by the judges become part of or comprise the complete questionnaire.

Probably the most commonly used method in attitudinal research is the Likert scale. This method generally makes use of a bipolar, five-point response range (i.e., strongly agree to strongly disagree). Questions that all respondents provide similar responses to are usually eliminated. The remaining questions are used to comprise the scale.

The last scale, the Guttman scale, requires that an attitudinal scale measure only one dimension or concept. The questions or statements must be progressive so that if the respondent answers positively to a question, he or she must respond the same to the following one.

Sampling

A population is the complete group or class from which information is to be gathered. A sample is a group chosen from within a target population to provide information sought.

Probability Theory

Probability theory is based on the concept that over time there is a statistical order in which things occur. The knowledge that over time things tend to adhere to a statistical order allows one to choose samples that are representative of a population in general.

Probability Sampling

The general goal when choosing a sample is to obtain one that is representative of the target population. Representation requires that every member in the population or the sampling frame have an equal chance of being selected for the sample. This is a probability sample. Four types of probability samples exist, as follows:

1. Simple random samples: A simple random sample is one in which all members of a given population have the same chance of being selected. Furthermore, the selection of each member must be independent from the selection of any other members.
2. Stratified random samples: These strata are selected based on specified characteristics that the researcher wishes to ensure for inclusion within the study. This type of sample requires the researcher to have knowledge of the sampling frame's demographic characteristics. These characteristics (selected variables) are then used to create the strata from which the sample is chosen.
3. Systematic samples: There seems to be some debate over this type of sampling. It includes random selection and initially allows inclusion of every member of the sampling frame. With a systematic sample, every nth item in the sampling frame is included in the sample.
4. Cluster samples: The last of the probability sampling methods is the cluster sample (also known as area "probability sample"). This sample consists of randomly selected groups, rather than individuals. The population to be surveyed is divided into clusters. Subsequent subsamples of the clusters are then selected.

Nonprobability Sampling

The major difference between probability and nonprobability sampling is that one provides the opportunity for all members of the sampling frame to be selected, whereas the other does not.

1. **Purposive samples:** Among the nonprobability samples, the purposive sample seems to be the most popular. Based on the researcher's skill, judgment, and needs, an appropriate sample is selected. When the subjects are selected in advance based on the researcher's view that they reflect normal or average scores this is sometimes referred to as "typical-case" sampling. If subgroups are sampled to permit comparisons among them, this technique is known as "stratified purposeful" sampling.

2. **Quota samples:** These types of research efforts often rely on quota sampling. For this type of sample, the proportions are based on the researcher's judgment for inclusion. Selection continues until enough individuals have been chosen to fill out the sample.

3. **Snowball samples:** Although not a highly promoted form of quantitative sampling, snowball sampling is commonly used as a qualitative technique. The snowball sample begins with a person or persons who provide names of other persons for the sample.

4. **Convenience sample:** Undoubtedly the last choice for a sample is the convenience sample or available subjects sample. Here there is no attempt to ensure any type of representativeness. Usually this sample is a very abstract representation of the population or target frame. Units or individuals are chosen simply because they were "in the right place at the right time." Convenience samples are often useful as explorations on which future research may be based.

Sample Size

Usually, the sample size is the result of several elements: how accurate the sample must be; economic feasibility (how much one has to spend); the availability of requisite variables (including any subcategories); and accessibility to the target population.

Confidence Levels

Addressing how large the sample should be requires an understanding of confidence intervals. The smaller the confidence interval the more accurate the estimated sample. The estimated probability that a population parameter will fall within a given confidence interval is known as the "confidence level." To reduce sampling error the researcher desires a smaller confidence interval. To do so, he or she selects a smaller confidence level.

Generally, in social science research, one seeks a sample size that varies 95 times out of 100 by 5% or less from the population. A sampling formula and sample size selection chart are provided in Chapter 9.

To ensure that the necessary numbers are obtained, it is recommended that one always oversample by 20%. If this is not enough, one can always add more observations as long as they are randomly selected from the same population and any time differences would not impact on responses.

Data Collection

There are four primary data collection techniques available: (1) survey; (2) interview (often classified in surveys); (3) observation; and (4) unobtrusive means. Probably the most frequently used method for data collection is the survey. There are two primary means for collecting data through surveys: mail surveys and self-completing questionnaires. A common means of distributing questionnaires is through the mail or by direct distribution.

Mail surveys allow for use of fairly large samples, broader area coverage, and the cost in terms of time and money can be kept to a minimum. Additional advantages include that no field staff is required, it eliminates the bias effect possible in interviews, allows the respondents greater privacy, places fewer time constraints on the respondents so that more consideration can be given to the answers, and offers the chance of a high percentage of returns improving the representativeness of the sample. The most frustrating disadvantage of mail surveys is the lack of responses.

Ways to increase response rates include offering some type of remuneration or "reward" for completing the survey; appealing to the respondent's altruistic side; using an attractive and shortened format; being able to indicate that the survey is sponsored or endorsed by a recognizable entity; personalizing the survey; and enhancing the timing of the survey.

The self-administered survey is generally a written questionnaire that is distributed to the selected sample in a structured environment. Respondents are allowed to complete the survey in a given time period and then return it to the researcher, often through an emissary of the sample.

There are several types of interviews that may be used in criminologic research. Probably the most used type of interview in criminal justice research is the structured interview. This requires the use of close-ended questions that every individual interviewed must be asked in the same order. Responses are set and can be checked off by the interviewer. It is in reality a questionnaire that is being administered orally with the interviewer completing the form for the respondent.

The unstructured interview offers respondents open-ended questions where no set response is provided. Although this allows for more in-depth responses, it is much more difficult to quantify the responses.

The in-depth interview can make use of both fixed and open-ended questions. The difference from the others is that the interviewer can further explore why the response was given and could ask additional qualifying questions.

Face-to-face interviews provide contact between the researcher and the respondent. This contact can be positive reinforcement for participating in the research. Often simply receiving a questionnaire in the mail can be very sterile, and that lack of personal touch might cause would-be respondents simply to ignore the survey.

The advantages of telephone surveying begin with the ability to eliminate a field staff and the creation of a very small in-house staff. It is also easier to monitor interviewer bias; specifically, it eliminates any nonverbal cues. In addition, the telephone interview is less expensive and very quick.

Like survey research, observational research has its advantages and disadvantages. One of the best advantages is the direct collection of the data. Rather than having to rely on what others have seen, here the researcher relies only on his or her observations.

The last means for collecting data is through what is referred to as "unobtrusive measures." These are any methods of data collection where the subjects of the research are completely unaware of being studied and that study is not observational. Two of the more commonly used types of unobtrusive data collection have been discussed previously in this chapter: the use of archival data (analyzing existing statistics or documents) and content analysis.

Data Processing

Data processing consists of data coding, data entry, and data cleaning. Coding is simply assigning values to the data for statistical analyses. Once the data are coded they can then be entered into a computer software program for analyses. The key to data entry is accuracy. Data cleaning is the preliminary analysis of your data. Here one "cleans" any mistakes that might have occurred during the initial recording of data or data entry.

Dealing with Missing Data

There are a number of ways to deal with missing information. If the data have been left out on a single question one may chose to input it as a nonresponse. Another option is to assume that the missing data is caused by an oversight rather than an intentional omission. Yet another option is to exclude the survey instrument containing the omission from the analysis. If the instrument used is a Likert scale the response may be classified as "neither agree nor disagree" or "do not know" (depending on how the scale is worded).

Recoding Data

By collapsing the categories into logical groupings, the data can be presented clearly and concisely. The resulting table could easily be collapsed even farther if necessary. In collapsing data, recall the earlier discussions of the different levels of measurement. Higher level data (e.g., ratio or interval data) can be collapsed down into lower level data (e.g., ordinal or nominal data) but one cannot do the opposite and convert lower level data to a higher level.

Data Analysis

There are three types of data analysis: (1) univariate analysis, (2) bivariate analysis, and (3) multivariate analysis. Univariate analysis is the examination of the distribution of cases on only one variable at a time. Bivariate analysis is when the relationship between two variables is examined. Multivariate analysis is the examination of three or more variables. This technique is inferential in nature in that one has usually already conducted both descriptive and comparative statistical analyses (through univariate and bivariate analyses) of the data and now seeks to examine the relationships among several variables.

Statistical Analysis

There are three types of statistics: (1) descriptive, whose function is describing the data; (2) comparative, whose function is to compare the attributes (or subgroups) of two variables; and (3) inferential, whose function is to make an inference, estimation or prediction about the data.

Frequency Distributions

There are four types of frequency distributions to be familiar with: (1) absolute, (2) relative, (3) cumulative, and (4) cumulative relative. The most common are absolute frequency distributions that display the raw numbers and relative frequency distributions that convert the numbers into percentages for easier interpretation by readers.

Displaying Frequencies

Frequencies, by whole numbers or percentages in table form, are just one means of describing the data. Other means include pie charts, bar graphs, histograms and polygons, line charts, and maps.

Other Ways to Describe the Data

In addition to frequency distributions the researcher may describe the properties of the data through measures of central tendency, measures of variability, skewness, and kurtosis. The three most common measures of central tendency are the mean, median, and mode. The mean is the arithmetic average. The median is the midpoint. The mode is the most frequently occurring number. The uses of measures of central tendency and how to compute them are discussed in detail in Chapter 11.

The three main measures of variability are range, variance, and standard deviation. The range is simply the difference between the highest and lowest scores. Variance is the difference between the scores and the mean. Standard deviation indicates how far from the mean is the score actually.

By using polygons, line charts, or scatterplots, researchers are able to see the distribution of the data. If it is a normal distribution, the researcher is able to use a broader range of statistical techniques. If it is a nonnormal (also known as nonparametric) distribution, the statistical techniques that may be used are more limited. Skewness alerts the researcher to the presence of outliers. Kurtosis refers to the amount of smoothness or pointedness of the curve. Such knowledge aids statisticians in conducting analyses of the data.

Inferential Statistics

Inferential statistics allow the researcher to develop inferences (predictions) about the data. By using statistical analyses researchers are able to make estimations or predictions about the data. If the sample is representative, these predictions may be extended to the population from which the data were drawn. Inferential statistics allow criminologic researchers to conduct research that can be generalized to larger populations within society.

Measures of Association

Measures of association are the means by which researchers are able to determine the strengths of relationships among the variables that are being studied. The measure of association that is used is dependent on the type of analysis that is being conducted, the distribution of the data, and the level of data that is used. There are many measures of association that are used in criminologic research. Individual researchers may prefer to use different measures. However, there are some measures that are considered to be standards. Lambda is commonly used for nominal level data. Gamma is commonly used for ordinal level research. Pearson r and R^2 are commonly used for interval and ratio level data.

Statistical Significance

Statistical significance is how researchers determine whether their sample findings are representative of the population that they are studying. If one is using a complete enumeration (the entire population is studied rather than just a sample from that population) then statistical significance is a moot point in that one already knows that the population is accurately represented. However, complete enumerations are rare in that very small populations limit generalizability and large populations are too costly and time-consuming to study as a whole.

Statistical significance is based on probability sampling. Generally, social science research desires a statistical significance of 0.05 or better. This means that one is 95% confident that the findings represent the population that was sampled. As seen in Chapter 12, determining statistical significance varies depending on the statistical procedures that are used.

Bivariate Analysis

Bivariate analysis is the examination of the relationship between two variables. Usually this involves attempting to determine how a dependent variable is influenced by an independent variable. The more commonly used bivariate techniques are crosstabulation (contingency tables); chi-square; and bivariate (simple linear) regression. In assessing the relationship between the two variables the contingency table or regression scatterplot results, the measure of association (e.g., gamma or R^2), and the level of statistical significance (it is hoped 0.05 or lower) are examined.

Multivariate Analysis

Multivariate analysis is the examination of the relationship between three or more variables. Usually this involves attempting to determine how a dependent variable is influenced by several independent variables. This methodology offers more insights than bivariate analysis in that one is able to study the relationships among several variables at one time. The more commonly used multivariate techniques are correlation, multiple regression, Student t test, ANOVA, discriminant analysis, probit regression, factor analysis, and path analysis. In assessing the relationship between the multiple variables one examines the correlation table, regression scatterplot results, or other indicators (depending on the technique used); the measure of association (e.g., Pearson r, R^2, Wilks lambda, or other appropriate measures); and the level of statistical significance.

Nonparametric Techniques

When the distribution of the data is not normal, standard statistical techniques are usually not appropriate. In those cases, other procedures that are used include chi-square, nonparametric correlation, and nonlinear regression. As noted in Chapter 12, the reader is advised to consult with a statistician in these circumstances.

Writing the Research

In Chapter 13 a research paper example was provided. The outline recommended in that chapter is as follows:

1. Title page: The title of the research paper (article or thesis) should tell the reader in clear and concise terms what the research is about.
2. Abstract: The abstract is the summary or synopsis of the information being presented in the paper and starts with the research title. The abstract presents the paper's major argument and describes the methods that were used.
3. Introduction: This section establishes the research problem and the literature that supports its existence and the reason to research it. In addition, this section reports the research question and hypothesis and briefly describes what was done.
4. Methodology: The methodology section is the mainstay of the paper. The writer should discuss the hypothesis, the research design, and the data gathering technique. This includes explaining the research population, the sampling frame, and the questionnaire or other method used to gather the data. It is in this section that the researcher can fully explain where, when, how, and why the data were attained and analyzed.
5. Results: In this section the writer describes the sample's characteristics, the statistical techniques used, the results, and whether they supported the hypothesis. It is within this section that various tables, graphs, and charts are commonly used to describe the data.
6. Conclusions: Usually the last section of the research paper is the conclusion. Generally, this section is used to offer insights about the research, whether it did what was expected, and possible problems. This section can also be used to discuss implications of the research and to provide a forum for suggestions for future related research.

7. References: When using other sources, whether quoting, paraphrasing, or simply as an outlet for affirming what is already known, the source should be recognized. This is done throughout the paper by various citation methods. The complete list of references or bibliography is provided at the end of the paper. The format depends on the source to which the paper is being submitted. Most papers written for criminal justice and criminology follow the APA style for reference citing in the text and at the end of the paper.

8. Tables and figures: Tables and figures are included in research papers to aid the author in presenting and explaining information. Regardless of their location, tables and figures need to be clear as to their content and readily understandable.

9. Appendixes: It is often useful to include a copy of the questionnaire or other tools that need to be shared but just do not belong in the body of the paper. There are no limitations to the number of appendices a paper can have except for those established by the instructor or the journal.

Summary

This chapter summarizes those issues from throughout the text that are important enough to warrant repetition. It is hoped that this text has provided the knowledge and insights necessary for successfully understanding and conducting research in criminal justice or criminology. It should be remembered that whether the research is applied or basic, qualitative or quantitative, the basic steps are applicable to each. There are five primary steps in conducting research:

1. Identifying the problem: Identifying or determining the problem, issue, or policy to be studied is what sets the groundwork for the rest of the research.

2. Research design: The research design is the "blueprint," which outlines how the research is conducted.

3. Data collection: Regardless of the research design, data collection is a key component. A variety of methods exist. They include surveys, interviews, observations, and previously existing data.

4. Data analysis: Proper analysis and interpretation of the data is integral to the research process.

5. Reporting: The last phase of any research project is the reporting of the findings. Regardless of the audience or the medium used, the findings must be coherent and understandable.

Adler, E. S., & Clark, R. (2007). *How it's done: An invitation to social research* (3rd ed.). Belmont, CA: Wadsworth Publishing.

Alarid, L. F. (2009). Risk factors for potential occupational exposure to HIV: A study of correctional officers. *Journal of Criminal Justice, 37*, 114–122.

Bachman, R., & Schutt, R. K. (2008). *Fundamentals of research in criminology and criminal justice.* Thousand Oaks, CA: Sage Publications.

Banyard, P., & Grayson, A. (2009). *Introducing psychological research* (3rd ed.). New York: Palgrave Macmillan.

Berg, B. (2008). Qualitative research methods for the social sciences (7th ed.). Columbus, OH: Allyn & Bacon.

Bergman, M. M. (2009). *Mixed methods research.* Thousand Oaks, CA: Sage Publications.

Bickman, L., & Rog, D. J. (2009). *The SAGE handbook of applied social science research methods* (2nd ed.). Thousand Oaks, CA: Sage Publications.

Bryman, A. (2008). *Social research methods.* New York: Oxford University Press.

Burns, M., & Dioquino, T. (1997). *Standardized field sobriety test battery.* Tallahassee, FL: Florida Department of Transportation.

Chappell, A. T. (2009). The philosophical versus actual adoption of community policing: A case study. *Criminal Justice Review, 34*(1), 5.

Charles, G. (2009). How spirituality is incorporated in police work: A qualitative study. *FBI Law Enforcement Bulletin, 78*(5), 22–25.

Colton, D., & Covert, R. W. (2007). *Designing and constructing instruments for social research and evaluation* (Research methods for the social sciences). San Francisco: John Wiley & Sons.

Creswell, J. W. (2008). *Research design: Qualitative, quantitative, and mixed methods approaches* (3rd ed.). Thousand Oaks, CA: Sage Publications.

Dantzker, M. L. (2010). *Psychologists' role and police pre-employment psychological screening.* Ann Arbor, MI: ProQuest Company.

Dantzker, M. L., & Ali-Jackson, N. (1998). Examining students' perceptions of policing and the effect of completing a police-related course. In M. L. Dantzker, A. J. Lurigio, M. J. Seng, & J. M. Sinacore (Eds.), *Practical applications for criminal justice statistics* (pp. 195–210). Boston: Butterworth-Heinemann.

Dantzker, M. L., & Eisenman, R. (2007). Sexual attitudes of criminal justice college students: Attitudes toward homosexuality, pornography, and other sexual matters. *American Journal of Psychological Research, 3*(1), 43–48.

Dantzker, M. L., & McCoy, J. H. (2006). Psychological screening of police recruits: A Texas perspective. *Journal of Police and Criminal Psychology, 21*(1), 23–32.

Dantzker, M. L., & Waters, J. E. (1999). Examining students' perceptions of policing: A pre- and post-comparison between students in criminal justice and non-criminal justice courses. In M. L. Dantzker (Ed.), *Readings for research methods in criminology and criminal justice* (pp. 27–36). Woburn, MA: Butter-worth-Heinemann.

DeMatteo, D., Marlowe, D. B., Festinger, D. S., & Arabia, P. L. (2009). Outcome trajectories in drug court: Do all participants have serious drug problems? *Criminal Justice and Behavior, 36*(4), 354–368.

Drake, B., & Jonson-Reid, M. (2008). *Social work research methods: From conceptualization to dissemination*. Columbus, OH: Allyn & Bacon.

Dunn, D. S. (2009). *Research methods for social psychology*. Sussex, England: John Wiley & Sons.

Eck, J. E., & La Vigne, N. G. (1994). Using Research: A primer for law enforcement managers. Washington, DC: Police Executive Research Forum.

Eisenman, R., & Dantzker, M. L. (2006). Gender and ethnic differences in sexual attitude at a Hispanic-serving university. *Journal of General Psychology, 133*(2), 153–162.

Flanyak, C. M. (1999). Accessing data: Procedures, practices, and problems of academic researchers. In M. L. Dantzker (Ed.), *Readings for research methods in criminology and criminal justice* (pp. 157–180). Woburn, MA: Butterworth-Heinemann.

Ford, M., & Williams, L. (1999). Human/cultural diversity training for justice personnel. In M. L. Dantzker (Ed.), *Readings for research methods in criminology and criminal justice* (pp. 37–60). Woburn, MA: Butterworth-Heinemann.

Fowler, F. J. (Ed.). (2009). *Survey research methods* (applied social research methods). Thousand Oaks, CA: SAGE Publications.

Frankfort-Nachmias, C., & Leon-Guerrero, A. (2008). *Social statistics for a diverse society*. New York: Pine Forge Press.

Gavin, H. (2008). *Understanding research methods and statistics in psychology*. Thousand Oaks, CA: SAGE Publications.

Gershon, R. R. M., Barocas, B., Canton, A. N., Li X., & Vlahov, D. (2009). Mental, physical, and behavioral outcomes associated with perceived work stress in police officers. *Criminal Justice and Behavior, 36*(3), 275–289.

Gillham, B. (2009). *Developing a questionnaire (real world research)* (2nd ed.). New York: Continuum International Publishing Group.

Given, L. M. (2008). *The SAGE encyclopedia of qualitative research methods*. Thousand Oaks, CA: Sage Publications.

Gover, A. R., Jennings, W. G., & Tewksbury, R. (2009). Adolescent male and female gang members' experiences with violent victimization, dating violence, and sexual assault. *American Journal of Criminal Justice, 34*(1/2), 103–118.

Hagan, F. E. (2006). *Essentials of research methods in criminal justice and criminology* (2nd ed.). Columbus, OH: Allyn & Bacon.

Harpster, T., Adams, S. H., & Jarvis, J. P. (2009). Analyzing 911 homicide calls for indicators of guilt or innocence: An exploratory analysis. *Homicide Studies, 13*(1), 69.

Hernandez, M. (2009). *Examining family factors and acculturation issues contributing to adolescent substance abuse in a Mexican American population*. Ann Arbor, MI: ProQuest Company.

Hilinski, C. M. (2009). Fear of crime among college students: A test of the shadow of sexual assault hypothesis. *American Journal of Criminal Justice, 34*(1/2), 84–103.

Hunter, R. D. (1988). *The effects of environmental factors upon convenience store robberies in Florida*. Tallahassee, FL: Florida Department of Legal Affairs.

Hunter, R. D. (1999). Officer opinions on police misconduct. *Journal of Contemporary Criminal Justice, 15*(2), 155–170.

Hunter, R. D., & Wood, R. L. (1994). Impact of felony sanctions: An analysis of weaponless assaults upon American police. *American Journal of Police, 12*(1), 65–89.

Janku, A. D., & Yan, J. (2009). Exploring patterns of court-ordered mental health services for juvenile offenders: Is there evidence of systemic bias? *Criminal Justice and Behavior, 36*(4), 402–419.

Jeffery, C. R. (1990). *Criminology: An interdisciplinary approach*. Englewood Cliffs, MA: Prentice Hall.

Kaplan, A. (1963). *The conduct of inquiry*. New York: Harper & Row, Publishers.

Keith, T. Z. (2005). *Multiple regression and beyond*. Columbus, OH: Allyn & Bacon.

Kline, R. B. (2009). Becoming a behavioral science researcher: A guide to producing research that matters. New York: The Guildford Press.

Klinger, D. A., & Brunson, R. K. (2009). Police officers' perceptual distortions during lethal force situations: Informing the reasonableness standard. *Criminology & Public Policy, 8*(1), 117.

Kraska, P. B., & Neuman, W. L. (2008). *Criminal justice and criminology research methods*. Columbus, OH: Allyn & Bacon.

Lavrakas, P. J. (2008). *Encyclopedia of survey research methods*. Thousand Oaks, CA: Sage Publications.

Lee, W-J., Joo, H-J., & Johnson, W. W. (2009). The effect of participatory management on internal stress, overall job satisfaction, and turnover intention among federal probation officers. *Federal Probation, 73*(1), 33–41.

Lowenkamp, C. T., Hubbard, D., Makarios, M. D., & Latessa, E. J. (2009). A quasi-experimental evaluation of thinking for a change: A "real-world" application. *Criminal Justice and Behavior, 36*(2), 137–149.

Lurigio, A. J., Greenleaf, R. G., & Flexon, J. L. (2009). The effects of race on relationships with the police: A survey of African American and Latino youths in Chicago. *Western Criminology Review, 10*(1), 29–41.

Maxfield, M. G., & Babbie, E. R. (2009). *Basics of research methods for criminal justice and criminology* (2nd ed.). Belmont, CA: Wadsworth Publishing.

McBurney, D. H., & White, T. L. (2007). *Research methods* (7th ed.). Belmont, CA: Wadsworth Publishing.

Memory, J. M. (1999). Some impressions from a qualitative study of implementation of community policing in North Carolina. In M. L. Dantzker (Ed.), *Readings for research methods in criminology and criminal justice* (pp. 1–14). Woburn, MA: Butterworth-Heinemann.

Monto, M. A., & Julka, D. (2009). Conceiving of sex as a commodity: A study of arrested customers of female street prostitutes. *Western Criminology Review, 10*(1), 1–13.

Morgan, S. E., Reichert, T., & Harrison, T. R. (2002). *From numbers to words: Reporting statistical results for the social sciences*. Columbus, OH: Allyn & Bacon.

Moriarty, L. J. (1999). The conceptualization and operationalization of the intervening dimensions of social disorganization. In M. L. Dantzker (Ed.), *Readings for research methods in criminology and criminal justice* (pp. 15–26). Woburn, MA: Butterworth-Heinemann.

Nachmias, D., & Nachmias, C. (2000). *Research methods in the social sciences* (6th ed.). New York: Worth Publishers.

Schultz, D. P., Hudak, E., & Alpert, G. P. (2009). Emergency driving and pursuits. *FBI Law Enforcement Bulletin, 78*(4), 1–6.

Senese, J. D. (1997). *Applied research methods in criminal justice.* Chicago: Nelson-Hall Publishers.

Shaughnessy, J. J., Zechmeister, E. B., & Zechmeister, J. S. (2008). *Research methods in psychology* (8th ed.). Dubuque, OH: McGraw-Hill Humanities/Social Sciences/Languages.

Stevens, D. J. (1999). Women offenders, drug addiction, and crime. In M. L. Dantzker (Ed.), *Readings for research methods in criminology and criminal justice* (pp. 61–74). Woburn, MA: Butterworth-Heinemann.

Talarico, S. M. (1980). *Criminal justice research: approaches, problems, and policy.* Cincinnati, OH: Anderson Publishing Company.

Vieira, T. A., Skilling, T. A., & Peterson-Badali, M. (2009). Matching court-ordered services with treatment needs: Predicting treatment success with young offenders. *Criminal Justice and Behavior, 36*(4), 385–401.

Violanti, J. M., Fekedulegn, D., Charles, L. E., Andrew, M. E., Hartley, T. A., Mnatsakanova, A., & Burchfiel, C. M. (2009). Suicide in police work: Exploring potential contributing influences. *American Journal of Criminal Justice, 34*(1/2), 41–46.

Vito, G. F., Kunselman, J. C., & Tewksbury, R. (2008). *Introduction to criminal justice research methods: An applied approach.* Springfield, IL: Charles C. Thomas.

Walker, J., & Maddan, S. (2009). *Statistics in criminology and criminal justice: Analysis and interpretation.* Sudbury, MA: Jones and Bartlett Publishers.

Wallace, W. (1971). *The logic of science in sociology.* New York: Aldine deGrutyer. Weisburd, D., & Britt, C. (2007). *Statistics in criminal justice* (3rd ed.). New York: Springer Science+Business Media LLC.

Wolfgang, M. E., Figlio, R. M., & Sellin, T. (1972). *Delinquency in a birth cohort.* Chicago: University of Chicago Press.

Research Ethics Review Application to the Walden University Institutional Review Board Requesting Approval to Conduct Research

❏ By checking this box, the submitter of this application is providing a digital signature confirming that she or he

    A. has read all of the instructions throughout the application;

    B. understands that failure to follow the instructions completely will result the application being returned to the submitter; and

    C. understands that noncompliance with IRB instructions and policies can result in consequences including but not limited to invalidation of data, revocation of IRB approval, and dismissal from Walden University.

Important Note for Student Researchers

It is the student's responsibility to make sure that the faculty-approved IRB application and all supporting materials are submitted to *IRB@waldenu.edu*. The IRB staff always confirms receipt of IRB materials and *a researcher may not begin recruiting participants or collecting data (including pilot data) until explicit IRB approval has been received from IRB@waldenu.edu.* Data collection that is begun prior to IRB approval does not qualify for academic credit toward degree requirements. Further, students collecting data without IRB approval risk expulsion.

What Is IRB Approval?

The Institutional Review Board (IRB) consists of staff and faculty members from each of Walden's major research areas and is responsible for ensuring that all Walden University research complies with the university's ethical standards as well as U.S. federal regulations and any applicable international guidelines. IRB approval indicates the institution's official assessment that the potential risks of the study are outweighed by the potential benefits. The IRB reviewers strive to limit their methodological comments to only those that impact either the risk or benefit level of the study, thus affecting the welfare of participants and stakeholders. IRB approval lasts for 1 year and may be renewed. Outside of the explicit dates and terms of IRB approval, researchers are not entitled to any protections, recognition, funding, or other support provided by Walden University or its affiliates. More detail about the IRB review process can be found at Walden's IRB Web site or by sending a specific request to IRB@waldenu.edu.

Who Should Use This IRB Application Form?

This application should be completed by all students and faculty members who are conducting research projects of any scope involving collection or analysis of data from living persons (whether from surveys, interviews, observation, student work, or records of any type). The only categories of research that do not need to be submitted for IRB approval are literature reviews, hypothetical research designs, and faculty projects that are completely independent of Walden affiliation, resources, participants, and funding. IRB approval for course-based research projects should be obtained by the faculty member who designs the course. Research projects conducted by fulltime employees of Walden or related organizations are also under the purview of the Walden IRB, as per federal regulations. Instead of completing this form, staff researchers should send an email inquiry to irb@waldenu.edu to initiate the IRB approval process for staff research.

When Should I Work On and Submit My IRB Application?

Questions about the IRB application and related materials may be submitted to *IRB@waldenu.edu* at any time. Non-doctoral IRB applications will be reviewed as soon as the application is complete.

For doctoral students, the IRB application itself will not be accepted until all of the following are complete: the proposal has been approved by the University Research Reviewer, the proposal oral conference has been held, and the student has received formal proposal approval notification from the Office of Student Research Support.

It is expected that doctoral students will review IRB requirements as they are writing the proposal and to that end, this IRB application can be used as a worksheet to help think through the ethical issues of data collection. However, the student must complete the IRB application again after proposal approval in order to address the details of the final, approved research design.

How Long Does IRB Review Take?

Researchers should allow a minimum of 4–6 weeks for IRB review (4 weeks for minimal risk studies and 6 weeks for studies involving vulnerable populations). This form takes 1–2 hours to complete, depending on the complexity of the study. Once the IRB staff confirms that the IRB application is complete, the IRB application will be scheduled for review at the next available IRB meeting (typically within 10 business days). Feedback from the board will be returned within 5 business days (amounting to a total of 15 business days for the initial review). Note that when a study is "approved with revisions" that the researcher should allow an additional 10–15 business days for those revisions to be reviewed and approved. If the revisions do not adequately address the ethical concerns, then an additional round of revisions and review might be necessary. The IRB members make every effort to make the required revisions as clear as possible.

Students should consult program guidelines and documents such as the dissertation guidebook in order to understand how long the proposal and IRB review steps will take and plan their study's timeline accordingly. Exceptions to approval procedures cannot be made in order to accommodate personal or external deadlines (e.g., limited access to participants).

Can I Contact My Research Participants Before IRB Approval?

Note that researchers may NOT begin recruiting participants (i.e., getting consent form signatures) prior to IRB approval. The only documents that may be signed before IRB approval are Data Use Agreements or Letters of Cooperation from community partners and Confidentiality Agreements that are signed by transcribers, statisticians, and research assistants who might have access to the raw data. If you have questions about who should sign what, please email *IRB@waldenu.edu* for help.

What If I Need to Change My Research Procedures after IRB Approval?

Researchers must resubmit any IRB materials relevant to the change, along with a Request for Change in Procedures form, which can be found on the Walden IRB Web site. As long as the proposed changes do not increase the level of risk, the request will be treated as an expedited review.

Overview of Requirements of This IRB Application

General Description of the Proposed Research

- Translate your research question(s) into lay language.
- Provide specific descriptions of the tasks the participants will be asked to complete.

Data Collection Tools

- Submit all documents and authorizations related to data collection including surveys, interview questions, evidence of compliance with copyright holder's terms of usage, permission to reproduce the instrument in the dissertation, or confirmation that the tool is public domain (as applicable).

Description of the Research Participants

- Describe the study population, particularly inclusion and exclusion criteria.
- If applicable, complete extra sections relevant to working with children, facility residents, or other protected populations.

Community Research Stakeholders and Partners

- Submit a signed Letter of Cooperation from any community partner who will be involved in identifying potential participants or collecting data.
- Submit a signed Data Use Agreement from any organization that will be providing records to the researcher.
- Describe your plan for sharing your research results with relevant stakeholders.

Potential Risks and Benefits

- Describe anticipated risks and benefits of study participation.
- Make provisions to minimize risks to research participants and document those procedures in this online application.

Data Integrity and Confidentiality

- Describe procedures to maintain data confidentiality and integrity.
- If data includes personal identifiers, submit signed certificates of confidentiality for everyone who has access to the data (except faculty members).
- If applicable, complete extra sections relevant to protected health information.

Potential Conflicts of Interest

- Disclose and manage potential conflicts of interest.

Informed Consent

- Make provisions to obtain and document informed consent from all study participants and the appropriate parents, guardians, or caregivers.
- Submit **unsigned** copies of any relevant consent documents.

Final Checklist and Electronic Signatures

- Students must obtain faculty approval (via electronic signature) before submitting this form to *IRB@waldenu.edu.*

This form must be completed and submitted via email. If you have questions as you are completing the form, please contact IRB@waldenu.edu.

PROJECT INFORMATION	
1. Researcher's name (must match university records)	___ M L Dantzker ___
2. Every researcher must submit a copy of their Human Research Protections training completion certificate with this application. A completion certificate is good for 3 years. Walden accepts Human Research Protections training certificates from NIH, NCI, or CITI, though NIH is most strongly recommended. The NIH training takes 1-1.5 hours.	Select year training was completed Indicate which research ethics training module was completed: ☒ National Institute of Health (NIH): 　　http://phrp.nihtraining.com ☐ Colllaborative Institutional Training Initiative (CITI): 　　http://www.citiprogram.org ☐ National Cancer Institute (NCI) ☐ Other research ethics training: _____
3. Researcher's email address	ml.dantzker@waldenu.edu
4. Project title	Psychologists' Role and Police Pre-Employment Psychological Screening
5. Research collaborators and roles If researcher is a student, please provide the name of the faculty member supervising this research (such as the committee chair).	Dr. Sandra Mahoney
6. Email address(es) of the supervising faculty member and any other co-researcher collaborators	smahoney@waldenu.edu
7. Walden program affiliation(s) of researcher: ☐ Education ☐ Engineering ☐ Health Services ☐ Human Services ☐ Management ☐ Nursing ☒ Psychology ☐ Public Health ☐ Public Policy and Administration	

PROJECT INFORMATION

8. Select the study type that best describes the IRB approval requested:

☐ Doctoral pilot study prior to proposal approval (must provide a rationale in question 10 for why a pilot study is necessary prior to proposal approval).
☒ Ph.D. Dissertation (may include a pilot but pilot steps must also be described in question 13)
☐ Ed.D. Doctoral Study (may include a pilot but pilot steps must also be described in question 13)
☐ Master's thesis
☐ KAM study
☐ Research for a course (specify course number: _____ and course enddate: _____)
☐ Faculty Research
☐ Other: _____

GENERAL DESCRIPTION OF THE PROPOSED RESEARCH

9. Please check all the data collection methods below that are part of this study. Ensure that all items checked here are explained in item #13.

☐ Interview
☐ Focus group
☒ Survey or assessment that is initiated by the researcher
☐ Survey or assessment that is routinely collected by the site
☐ Analysis of student work products
☐ Analysis of existing public records or documents
☐ Analysis of existing privately held records (such as business records, school records) or documents
☐ Observation of people in public places
☐ Observation of people in school, workplace, or other non-public location
☐ Collection of physical specimens (e.g. blood, saliva)
☐ Other (please specify) _____

10. Using layperson terms please state your research question.	Whether there are differences between police psychologists and general clinical psychologists in the evaluative instruments or protocols used in conducting pre-employment screening for potential police officers and do they select those instruments or protocols for reasons of job-specific validity?
11. <u>Quantitative researchers</u>: Please list each variable of interest (identifying each, if applicable, as independent, dependent, or covariate) and briefly explain how they will be measured. <u>Qualitative researchers</u>: Please describe the phenomenon of interest and how it will be recorded. A subsequent section will ask for more detailed information about your data collection tools.	independent (psychologist's role) and dependent variables [items used and reason(s)] both will be measured through a self-administered questionnaire

GENERAL DESCRIPTION OF THE PROPOSED RESEARCH

12. Please briefly describe the analyses planned. Describe which statistical or analytical methods you will use to reveal expected relationships, differences, or patterns. The IRB is obligated to factor the rigor of the research design into the overall assessment of the potential risks and benefits of this study.	Data will be analyzed statistically for descriptors of all the variables using frequencies and Chi square. The comparative statistics between the independent (psychologist's role) and dependent variables [items used and reason(s)] will include Pearson's r, ANOVA, and multiple regression.

13. Please describe the research steps in enough detail that privacy and safety risks can be ascertained.

You must describe any of the following interactions with participants that apply to your study:
-Identification of potential participants
-Initial contact with potential participants
-Informed consent procedures
-If applicable, pilot activities (If changes need to be made based on the pilot, you will need to submit a
     Request for Change in Procedures form, which is found on the IRB website.)
-Data collection (examination of records, surveys, interviews, assessments, observations)
-Any intervention/treatment activities that are critical to the study even if provided by another entity
-Meeting again with participants to review interview transcripts and/or do membercheck (checking validity of
     findings)
-Dissemination of study's results to participants and stakeholders
If there are more than 10 steps, then you may email an attachment listing additional steps. Note that missing
     information will delay the IRB review.

	Detailed description	Duration of Data Collect Sessions	Exact Location	Communication Format (e.g., email, phone, in person, internet, etc.)
Step 1	No actual physical contact will be made. All contact will be through the U.S. Mail system and the Internet	2-4 weeks	Psychologists stratified and randomly selected throughout the country and members of the IACP Police Psychology Section.	U.S. Mail by letter; e-mail and Internet
Step 2	Completion of the questionnaire will represent consent.			
Step 3	Particpants will be stratified and randomly selected through an APA database and all members of the IACP's Police Psychology Section.	2-4 weeks		
Step 4	Letters and e-mails will be sent to participants w/hyperlink to questionnaire.	1-2 weeks		

GENERAL DESCRIPTION OF THE PROPOSED RESEARCH

Step 5	After 2 weeks a reminder post card nad/or e-mail will be sent to participants.			
Step 6	At the end of 8 weeks data will be reviewed			
Step 7	Data Analyses	1-2 weeks		
Step 8	Write - up	4-6 weeks		
Step 9				
Step 10				

Data Collection Tools

In order to approve your study, the IRB needs to *review the full text of each data collection tool* (e.g., surveys, interview questions, etc.). This application's final checklist will direct you to send your data collection tools and evidence of compliance with the copyright holder's usage terms at the same time you submit this IRB form. If any further changes are made to the data collection tools after they have been IRB-approved, you must submit those changes for IRB approval.

Read This If You Are Using a Published Instrument

Many assessment instruments published in journals can be used in research as long as commercial gain is not sought and proper credit is given to the original source (United States Code, 17USC107). However, publication of an assessment tool's results in a journal does not necessarily indicate that the tool is in the public domain.

The copyright holder of each assessment determines whether permission and payment are necessary for use of that assessment tool. Note that the copyright holder could be either the publisher or the author or another entity (such as the Myers and Briggs Foundation, which holds the copyright to the popular Myers-Briggs personality assessment). *The researcher is responsible for identifying and contacting the copyright holder* to determine which of the following are required for legal usage of the instrument: purchasing legal copies, purchasing a manual, purchasing scoring tools, obtaining written permission, obtaining explicit permission to reproduce the instrument in my dissertation, or simply confirming that the tool is public domain.

Even for public domain instruments, Walden University requires students to provide the professional courtesy of *notifying the primary author of your plan* to use that tool in your own research. Sometimes this is not possible, but at least three attempts should be made to contact the author at his or her most recently listed institution across a reasonable time period (such as 2 weeks). The author typically provides helpful updates or usage tips and asks to receive a copy of the results.

Many psychological assessments are restricted for use only by *suitably qualified individuals*. Researchers must check with the test's publisher to make sure that they are qualified to administer and interpret any particular assessments that they wish to use.

Read This If You Are Creating Your Own Instrument or Modifying an Existing Instrument

It is not acceptable to modify assessment tools without explicitly citing the original work and detailing the precise nature of the revisions. Note that even slight modifications to items or instructions threaten the reliability and validity of the tool and make comparisons to other research findings difficult, if not impossible. Therefore, unless a purpose of the study is to compare the validity and reliability of a revised measure with that of one that has already been validated, changes should not be made to existing measures. If the study is being conducted for the purpose of assessing the validity/reliability of a modified version of an existing measure, the original measure must also be administered to participants.

GENERAL DESCRIPTION OF THE PROPOSED RESEARCH

14. Are any of your data collection tools published?

☒ No.

☐ Yes, the following instrument is published: <u>(Insert name of instrument)</u>. I have consulted the copyright holder, <u>(Insert copyright holder name)</u>, and I have complied with all of the copyright holder's legal usage terms which include (check all that apply):

      ☐ Obtaining legal copies of the instrument

      ☐ Obtaining a legal copy of the manual or scoring kit

      ☐ Obtaining written permission to use the instrument in my research
        (submitted with this application)

      ☐ Obtaining explicit permission to reproduce the instrument in my dissertation
        (submitted with this application)

      ☐ Citation of source confirming that the tool is public domain: <u>(Insert citation here)</u>.

      ☐ Other: _____

If you are working with multiple copyright holders for different instruments, you must list the legal usage requirements for each additional instrument, following the format above:
<u>(Insert name(s) of additional tools and a list of the copyright holders' requirements for legal usage)</u>.

---

15. Did you create any of your data collection tools yourself?

☐ No.

☒ Yes, I created the following data collection tools: <u>(Pre-Screening of Police/Law Enforcement Candidates Psychologist Survey.</u>

Are you modifying an existing tool?

☒ No.

☐ Yes, the APA style citation for the original tool is _____; include my modifications _____; and these modifications are necessary because _____.

GENERAL DESCRIPTION OF THE PROPOSED RESEARCH

If you checked yes to either of the questions above, please answer the following:

Did an expert panel outside of the faculty committee review the tool(s)?
☐ No.
☒ Yes.
(Expert panel review is not required but increases validity of a student-designed tool and thus, factors into the ratio of benefits to risks.)

Were any of these tools piloted already in a previous, IRB-approved study?
☒ No.
☐ Yes. The IRB approval number was _____.
(Piloting is not required but factors into benefits/risks assessment.)

Do you plan to pilot any of these tools or procedures?
☒ No.
☐ Yes. (Briefly describe exactly what aspect of the study will be piloted and ensure the pilot steps are included in your response to question #13.)

DESCRIPTION OF THE RESEARCH PARTICIPANTS

16. Provide the target number of participants, including numbers per group if your study involves multiple groups or a separate pilot sample: 1160; 1000 stratified and randomly chosen through APA database, and 160 from IACP Clinical Psychology Section

Provide a brief rationale for this sample size: The proposed study will meet an alpha or confidence level of .05. Because the proposed study looks to use a quality sample, with less than a 5% error tolerance, it is seeking a sample size between 385 and 625. To reach this, 1160 individuals will be chosen for the sample.

17. Describe the population from which the sample will be chosen: Licensed clinical psychologists who are members of APA and/or the IACP Police Psychology Section

Give the population's approximate size: The sample is to be drawn from the APA group, which at the end of 2007 had over 98,000 members, 43,159 of whom were listed as clinical psychologists (research.apa.org). The entire population of the IACP group (160) is to be surveyed, too.

Describe how the research invitations will be managed to facilitate recruitment of an appropriate sample. (If you are using any quantitative methods, please describe what measures will be taken to obtain a sample that is as representative of the population as possible in terms of gender, ethnicity, or any other relevant demographics.): Members from the APA group will receive a form letter with the relevant information while members of the IACP will receive an e-mail invitation.

18. Now please list all criteria for inclusion and exclusion of participants in this study (such as relevant experiences, age, health conditions, etc). Your inclusion criteria should define all critical characteristics of your sample. Once you've defined inclusion criteria, if you have no further limitations on who can participate, just indicate "none" under exclusion criteria.

Inclusion criteria: Must be a licensed clinical psychologist
Describe how you will identify individuals who meet the inclusion criteria: clinical psychologist members of both groups must be licensed to be a member

Exclusion criteria: Must be a member of APA or the IACP Clinical Psychology Section
Describe how you will identify which individuals must be excluded: only the above will be contacted

DESCRIPTION OF THE RESEARCH PARTICIPANTS

19.   The checklist of vulnerable groups below will help you check your responses to questions 16-18 for potential ethical problems. The ethical challenge is to achieve the goal of equitable sampling that is appropriate to the research question while excluding vulnerable individuals whom the research procedures cannot adequately protect.

The potentially vulnerable populations listed below may only be specifically <u>recruited</u> when (a) the vulnerability status is directly related to the research question and (b) adequate measures are taken to ensure safety and voluntary participation.

<u>For each of the vulnerable groups below, indicate whether your articulated procedures will recruit any of the following as participants. You need to check ONE of the four boxes for each lettered category of vulnerable participants and add descriptions as indicated.</u>

A.   Minors (17 and under):
- ☐   Yes: I will be specifically recruiting minors as participants.
        <u>Describe how you will protect minors from pressure to participate.</u>
        <u>Describe how you will protect minors from safety and privacy risks.</u>
- ☐   Possible: My participants might be minors but I may not know if they are.
        <u>Describe how you will protect minors from pressure to participate.</u>
        <u>Describe how you will protect minors from safety and privacy risks.</u>
- ☐   No: I will screen age so I can exclude minors from my sample.
        <u>Explain which recruitment or screening procedure will enable exclusion.</u>
- ☒   No: My recruitment methods automatically exclude minors.

B.   Students of the researcher:
- ☐   Yes: I will be specifically recruiting my students as participants.
        <u>Describe how you will protect your students from pressure to participate.</u>
        <u>Describe how you will protect your students from safety and privacy risks.</u>
- ☐   Possible: My participants might be my students but I may not know if they are.
        <u>Describe how you will protect your students from pressure to participate.</u>
        <u>Describe how you will protect your students from safety and privacy risks.</u>
- ☐   No: I will screen student status so I can exclude my students from my sample.
        <u>Explain which recruitment or screening procedure will enable exclusion.</u>
- ☒   No: My recruitment methods automatically exclude my students.

C.   Subordinates of the researcher:
- ☐   Yes: I will be specifically recruiting my subordinates as participants.
        <u>Describe how you will protect your subordinates from pressure to participate.</u>
        <u>Describe how you will protect your subordinates from safety and privacy risks.</u>
- ☐   Possible: My participants might be my subordinates but I may not know if they are.
        <u>Describe how you will protect your subordinates from pressure to participate.</u>
        <u>Describe how you will protect your subordinates from safety and privacy risks.</u>
- ☐   No: I will screen subordinate status so I can exclude them from my sample.
        <u>Explain which recruitment or screening procedure will enable exclusion.</u>
- ☒   No: My recruitment methods automatically exclude my subordinates.

D.   Clients or potential clients of the researcher.
- ☐   Yes: I will be specifically recruiting my clients as participants:
        <u>Describe how you will protect clients from pressure to participate.</u>
        <u>Describe how you will protect clients from safety and privacy risks.</u>
- ☐   Possible: My participants might be my clients but I may not know if they are.
        <u>Describe how you will protect clients from pressure to participate.</u>
        <u>Describe how you will protect clients from safety and privacy risks.</u>

DESCRIPTION OF THE RESEARCH PARTICIPANTS

☐ No: I will screen my client status so I can exclude them from my sample.
Explain which recruitment or screening procedure will enable exclusion.
☒ No: My recruitment methods automatically exclude my clients.

E. Residents of any facility (nursing home, prison, assisted living, group home for minors):
☐ Yes: I will be specifically recruiting facility residents as participants.
Describe how you will protect facility residents from pressure to participate.
Describe how you will protect facility residents from safety and privacy risks.
☐ Possible: My participants might be facility residents but I may not know if they are.
Describe how you will protect facility residents from pressure to participate.
Describe how you will protect facility residents from safety and privacy risks.
☐ No: I will screen facility resident status so I can exclude them from my sample.
Explain which recruitment or screening procedure will enable exclusion.
☒ No: My recruitment methods automatically exclude facility residents.

F. Mentally/emotionally disabled individuals:
☐ Yes: I will be specifically recruiting mentally/emotionally disabled individuals as participants.
Describe how you will protect mentally/emotionally disabled individuals from pressure to participate.
Describe how you will protect mentally/emotionally disabled individuals from safety and privacy risks.
☐ Possible: My participants might be mentally/emotionally disabled but I may not know if they are.
Describe how you will protect mentally/emotionally disabled individuals from pressure to participate.
Describe how you will protect mentally/emotionally disabled individuals from safety and privacy risks.
☐ No: I will screen mentally/emotionally disabled status so I can exclude them from my sample.
Explain which recruitment or screening procedure will enable exclusion.
☒ No: My recruitment methods automatically exclude mentally/emotionally disabled individuals.

G. Individuals who might be less than fluent in English:
☐ Yes: I will be specifically recruiting non-English speakers as participants.
Describe how you will protect non-English speakers from pressure to participate.
Describe how you will protect non-English speakers from safety and privacy risks.
☐ Possible: My participants might be less than fluent in English but I may not know if they are.
Describe how you will protect non-English speakers from pressure to participate.
Describe how you will protect non-English speakers from safety and privacy risks.
☐ No: I will screen non-English speakers so I can exclude them from my sample.
Explain which recruitment or screening procedure will enable exclusion.
☒ No: My recruitment methods automatically exclude non-English speakers.

H. Elderly individuals (65+):
☐ Yes: I will be specifically recruiting elderly individuals as participants.
Describe how you will protect elderly individuals from pressure to participate.
Describe how you will protect elderly individuals from safety and privacy risks.
☒ Possible: My participants might be elderly but I may not know if they are.
Age is not a factor in that the individual can choose to participant without any knowledge of the researcher whether s/he chose not to because of age (random selection and no identifiers are sought that would lead to indentification of a specific individual.

☐ No: I will screen age so I can exclude elderly individuals from my sample.
Explain which recruitment or screening procedure will enable exclusion.
☐ No: My recruitment methods automatically exclude elderly individuals.

DESCRIPTION OF THE RESEARCH PARTICIPANTS

I. Individuals who are in crisis (such as natural disaster victims or persons with an acute illness):
- ☐ Yes: I will be specifically recruiting individuals in crisis as participants.
    - Describe how you will protect individuals in crisis from pressure to participate.
    - Describe how you will protect individuals in crisis from safety and privacy risks.
- ☐ Possible: My participants might be in crisis but I may not know if they are.
    - Describe how you will protect individuals in crisis from pressure to participate.
    - Describe how you will protect individuals in crisis from safety and privacy risks.
- ☐ No: I will screen crisis status so I can exclude them from my sample.
    - Explain which recruitment or screening procedure will enable exclusion.
- ☒ No: My recruitment methods automatically exclude individuals in crisis.

J. Economically disadvantaged individuals:
- ☐ Yes: I will be specifically recruiting economically disadvantaged individuals as participants.
    - Describe how you will protect economically disadvantaged individuals from pressure to participate.
    - Describe how you will protect economically disadvantaged individuals from safety and privacy risks.
- ☐ Possible: My participants might be economically disadvantaged but I may not know if they are.
    - Describe how you will protect economically disadvantaged individuals from pressure to participate.
    - Describe how you will protect economically disadvantaged individuals from safety and privacy risks.
- ☐ No: I will screen economic status so I can exclude them from my sample.
    - Explain which recruitment or screening procedure will enable exclusion.
- ☒ No: My recruitment methods automatically exclude economically disadvantaged individuals.

K. Pregnant women:
- ☐ Yes: I will be specifically recruiting pregnant women as participants.
    - Describe how you will protect pregnant women from safety and privacy risks.
- ☒ Possible: My participants might be pregnant but I may not know if they are.
    - Pregnancy is not a factor in that the individual can choose to participant without any knowledge of the researcher whether she chose not to because of pregnancy (random selection and no identifiers are sought that would lead to indentification of a specific individual.
- ☐ No: I will screen pregnancy status so I can exclude them from my sample.
    - Explain which recruitment or screening procedure will enable exclusion.
- ☐ No: My recruitment methods automatically exclude pregnant women.

L. Other vulnerable population: _____   ☐ Yes: I will be specifically recruiting another vulnerable group as participants.
    - Describe how you will protect this vulnerable group from pressure to participate.
    - Describe how you will protect this vulnerable group from safety and privacy risks.
- ☐ Possible: My participants might be part of this other vulnerable group but I may not know if they are.
    - Describe how you can minimize pressure to participate.
    - Describe how you can minimize safety/privacy risks that might particularly impact this group.
- ☐ No: I will screen this status so I can exclude them from my sample.
    - Explain which recruitment or screening procedure will enable exclusion.
- ☒ No: My recruitment methods automatically exclude this group.

DESCRIPTION OF THE RESEARCH PARTICIPANTS

20. Please briefly justify the inclusion of each vulnerable group for whom you answered "Yes" or "Possible" above in item 19. Ensure that this response provides a rationale for why it is impossible or unethical to conduct the research without the use of the protected population.	Inclusion of elderly and/or pregnant women is considered a possible given since individuals are not being screened out for either factor and may possibly be part of sample being randomly selected.
21. If competency to provide consent could possibly be an issue for any participants, describe how competency will be determined and your plan for obtaining consent. If not applicable, please indicate NA.	N/A

ADDITIONAL ISSUES TO ADDRESS WHEN PARTICIPANTS INCLUDE CHILDREN

(as per Federal Regulations)

22. Will your sample include individuals less than 18 years of age?
☐ Yes → Please complete questions 23-24.
☒ No → Please skip ahead to question 25.

23. If this study proposes to include minors, this inclusion must meet one of the following criteria for risk/benefit assessment, according to the federal regulations (link provided on the Walden IRB Web site).

Check the one appropriate box:
☐ Minimal risk.
☐ Greater than minimal risk, but holds prospect of direct benefit to participants.
☐ Greater than minimal risk, no prospect of direct benefit to participants, but likely to yield generalizable knowledge about the participant's disorder or condition.

24. Please explain how the criterion in question 25 is met for this study.	_____

ADDITIONAL ISSUES TO ADDRESS WHEN PARTICIPANTS INCLUDE PRISONERS

(as per Federal Regulations)

25. Is it possible that your sample will include prisoners?
☐ Yes → Please complete question 26.
☒ No → Please skip ahead to question 27.

ADDITIONAL ISSUES TO ADDRESS WHEN PARTICIPANTS INCLUDE PRISONERS
(as per Federal Regulations)

26. Enrollment of prisoners requires that the IRB is able to document that the seven conditions under federal regulations 45 CFR 46 Subpart C are met. If you plan to recruit individuals who are at high risk of becoming incarcerated in a penal institution during the research (e.g., participants with substance abuse history, repeat offenders, etc.), it is best that the IRB can address the Subpart C requirements at the time of initial review. Otherwise, if a participant becomes incarcerated during the course of the research and the IRB has not previously reviewed and approved your research for enrollment of prisoners, all research activity must immediately cease for that individual until review and application of Subpart C regulations occurs by the IRB.

A. Will this study examine the possible causes, effects, or processes of incarceration?
☐ Yes.
☐ No.

B. Will this study examine the facility as an institutional structure?
☐ Yes.
☐ No.

C. Will this study specifically examine the experience of being incarcerated?
☐ Yes.
☐ No.

D. Will this study examine a condition(s) particularly affecting these prisoners?
☐ Yes.
☐ No.

E. Will this study examine a procedure, innovative or accepted, that will have the intent or reasonable probability of improving the health or well being of the participants?
☐ Yes, and residents will be assigned to groups by _____.
☐ No.

Community Research Stakeholders and Partners

Research participants are individuals who provide private data through any type of interaction, whether verbal, observed, typed, recorded, written, or otherwise assessed. Research participants' understanding of the study and willingness to engage in research must be documented with CONSENT FORMS, *after* IRB approval. For example, an educator comparing two instructional strategies by interviewing adult students in his classes would need to have each participant student sign a consent form.

Community partners include any schools, classroom teachers, clinics, businesses, non-profits, government entities, residential facilities, or other organizations who are involved in your research project. Community partners' understanding of the study and willingness to engage in research must be documented with a LETTER OF COOPERATION. To continue with the same example, the educator comparing two instructional strategies would need a Letter of Cooperation from the school confirming (a) that the school approves the teacher's implementation of two different instructional strategies and (b) that the school approves the interview activities. In some cases a community partner will only provide a letter of cooperation after Walden has "officially" approved the research proposal. If this is the case, then enter a brief explanation of your planned steps in item 28. If you have questions about whether an individual or an organization should provide permission for some aspect of the research, please email IRB@waldenu.edu.

If a community partner's engagement in the research involves *providing any type of non-public records*, the terms of sharing those records must be documented in a DATA USE AGREEMENT, *before* IRB approval. Again using the same example, the educator comparing two instructional strategies will need a Data Use Agreement if he wants to analyze these students' past academic records or work products as part of the study. Data Use Agreements must be FERPA-compliant and HIPAA-compliant, as applicable to the setting.

A sample letter of cooperation and sample data use agreement can be downloaded from the IRB section of the Walden Web site. This IRB application's final checklist will direct you to email or fax your community partners' Letters of Cooperation and any applicable Data Use Agreements at the same time you submit this IRB form.

Stakeholders include the informal networks of individuals who would potentially be impacted by the research activities or results (such as parents, community leaders, etc). Walden students are required to disseminate their research results in a responsible, respectful manner and are encouraged to develop this dissemination plan in consultation with the relevant community partners. Sometimes it is appropriate to provide a debriefing session/handout to individual participants immediately after data collection in addition to a general stakeholders' debriefing after data analysis.

COMMUNITY RESEARCH STAKEHOLDERS AND PARTNERS

27. Please identify all community stakeholders who should hear about your research results and indicate your specific plan for disseminating your results in an appropriate format.	I intend to attempt to publish the results in a relevant journal such as the Journal of Police and Criminal Psychologists, Psychological Services, and/or The Counseling Psychologist..
28. Please specify the names and roles of any community partner organizations you propose to involve in identifying potential participants or collecting data. For each organization, identify the individual who is authorized to sign the Letter of Cooperation and any applicable Data Use Agreement (see definitions above). If you have no community research partner, that means you are solely relying on **public** records to recruit participants and collect data. This is also the place to enter any special notes about the timing of obtaining your letter of cooperation (e.g., if the partner will only sign a letter of cooperation AFTER IRB approval).	Kim Kohlhepp Staff Liaison, Psychological Services Section International Association of Chiefs of Police 515 North Washington Street Alexandria, VA 22314-2357 If you are planning on using the Walden Participant Pool, please check one of the following options: ☐ Notification that my study is eligible for placement on the website was granted from the Walden representative on Insert Date. ☐ My study needs to be reviewed by the Walden representative to determine its eligibility for placement on the site. (Please note, the IRB will coordinate with the Walden representative to make this determination.)
29. Please briefly describe how you chose each of the partners listed above.	Due to familiarity with the organization.

POTENTIAL RISKS AND BENEFITS

30. For each of the categories A-J below, carefully estimate risk level and describe the circumstances that could contribute to that type of negative outcome for **participants or stakeholders**. Minimal risk is acceptable but must be identified upfront. Minimal risk is defined as follows in U.S. federal regulations: "that the probability and magnitude of harm or discomfort anticipated in the research are not greater in and of themselves than those ordinarily encountered in daily life or during the performance of routine physical or psychological examinations or tests." Substantial risk is acceptable as long as adequate preventive protections are in place (which you will describe in item 31).

	Level of risk: check one	Description of risk: List the circumstances that could cause this outcome
A. Unintended disclosure of confidential information (such as educational or medical records)	☒ Not applicable ☐ Minimal risk ☐ Substantial risk	_____
B. Psychological stress greater than what one would experience in daily life (e.g., materials or topics that could be considered sensitive, offensive, threatening, degrading)	☒ Not applicable ☐ Minimal risk ☐ Substantial risk	_____
C. Attention to personal information that is irrelevant to the study (i.e., related to sexual practices, family history, substance use, illegal behavior, medical or mental health)	☒ Not applicable ☐ Minimal risk ☐ Substantial risk	_____
D. Unwanted solicitation, intrusion, or observation in public places	☒ Not applicable ☐ Not applicable ☐ Substantial risk	_____

POTENTIAL RISKS AND BENEFITS

E. Unwanted intrusion of privacy of others not involved in study (e.g. participant's family).	☒ Not applicable ☐ Minimal risk ☐ Substantial risk	——————
F. Social or economic loss (i.e., collecting data that could be damaging to any participants' or stakeholders' financial standing, employability or reputation)	☒ Not applicable ☐ Minimal risk ☐ Substantial risk	——————
G. Perceived coercion to participate due to any existing or expected relationship between the participant and the researcher (or any entity that the researcher might be perceived to represent)	☒ Not applicable ☐ Minimal risk ☐ Substantial risk	——————
H. Misunderstanding as a result of experimental deception (such as placebo treatment or use of confederate research assistants posing as someone else)	☒ Not applicable ☐ Minimal risk ☐ Substantial risk	——————
I. Minor negative effects on participants' or stakeholders' health (no risk of serious injury)	☒ Not applicable ☐ Minimal risk ☐ Substantial risk	——————
J. Major negative effects on participants' or stakeholders' health (risk of serious injury)	☒ Not applicable ☐ Minimal risk ☐ Substantial risk	——————

31. Explain what steps will be taken to minimize risks and to protect participants' and stakeholders' welfare. If the research will include protected populations, identify each group and answer this question for each group.	No personal data is being collected that would put subjects at any risk. Furthermore, no personal contact will be made. Everything is based solely on response to a questionnaire that requires no identifying data.
32. Describe the anticipated benefits of this research, if any, for individual participants.	Minimal unless they are specifically engaged in Police recruitment screening.
33. Describe the anticipated benefits of this research for society.	Will help in determining whether there is a difference among psychologists regarding psychological protocols for screening police candidates, to help establish whether there should be a difference, and to possibly assist in proposing a standardization of protocols which would add consistency to the screening process which does not currently exist.

DATA INTEGRITY AND CONFIDENTIALITY

34. In what format(s) will you obtain and subsequently store the data? (e.g., paper, electronic media, video, audio)	electronic (computer).
Where will you store the data?	Data will be stored on devices maintained by the researcher.
35. Describe what security provisions will be taken to protect this data during initial data collection, data transfer, and archiving. (e.g., privacy envelopes, password protection, locks)	Password protected
36. Describe what types of checks are in place to facilitate accuracy of data collection. Please note that the university's Office of Research Integrity and Compliance can audit the complete set of raw data at any time after IRB approval.	Only the sample will be provided access to the questionnaire. Otherwise, there are no other specific means to facilitate accuracy.
37. Explain exactly when and how the data disposal will occur. (Keeping raw data for five years is the minimum requirement).	At this point will keep data for a Five year minimum at which time all data files will be erased.
38. Describe the specific plans for handling adverse events involving research participants that might require immediate referral, stopping data collection, management of a new conflict of interest, reassessment of risks and benefits, or responding to breached confidentiality. These plans must be tailored by the researcher for the specific research context and population.	Using a Stratified Random selection for APA members who will be sent a letter directly from researcher and IACP section members will receive an e-mail from Researcher through organization's e-mail system should limit any adverse events. Furthermore, the general nature of the study as a self-administered questionnaire in itself should limit any adverse conditions.

39. Understanding the difference between confidentiality and anonymity:
Anonymous data contains absolutely zero identifiers and makes it impossible to determine who participated and who did not.
Confidential data contains one or more identifiers, but identifiers are kept private by the researcher. In order to protect participant privacy and assure that study participation is truly voluntary, anonymous data collection is preferred, whenever possible.

Is it possible to collect your data anonymously?
☐ No, my communications with potential participants and/or consent procedures require one or more of their identifiers (such as name, email address, or phone number) to be shared with me. But I confirm that I will provide complete confidentiality.
☒ Yes, I have designed my anonymous consent and data collection procedures so that identities are completely protected even from me, the researcher.

DATA INTEGRITY AND CONFIDENTIALITY

40. Will you retain a link between study code numbers and direct identifiers after the data collection is complete?
☒ No.
☐ Yes, but only to identify those participants who indicate that they want their data withdrawn.
☐ Yes, it is otherwise necessary because _____

41. Will you provide an identifier or potentially identifying link to anyone else besides yourself? ☒ No. ☐ Yes, it is necessary because _____.	_____
42. Explain who will approach potential participants to take part in the research study and what will be done to protect individuals' privacy in this process.	Members of the APA sample will receive a letter directly from the researcher with information for participation while members of the IACP group will receive an e-mail from an organization liaison. There will be no record kept of who particpated by any type of identifier.
43. Please list all individuals who will have access to the data (including research assistants, transcribers, statisticians, etc.). If you are a student, the IRB assumes that your supervising faculty members will have access to the data, so you do not need to list them.	Student researcher and supervising faculty

44. To ensure data confidentiality among your research colleagues, you will either need to obtain a signed Confidentiality Agreement for each person you listed for Question 43 or de-identify the data (by removing all identifying links) before anyone else has access to it. Please visit the IRB Web site to download a sample Confidentiality Agreement. This application's final checklist will direct you to send the IRB your signed Confidentiality Agreement(s) at the same time you submit this IRB form.

Please check all that apply.

☐ I will be emailing the signed confidentiality agreement(s) to IRB@waldenu.edu.
☐ I will be faxing the signed confidentiality agreement(s) to (626) 605-0472.
☐ Not applicable because I am the only one who will have access to the raw data.
☒ Not applicable because the accessible data is anonymous or de-identified.

45. Please confirm that you have made yourself aware of any state laws that might be relevant to this study's data collection activities including mandated reporting, privacy laws, etc. Please provide a plan for complying with these laws.	I am aware of all state laws and ethical considerations regarding this study's data collection. The process I have proposed to follow meets all laws and criteria.

ADDITIONAL ISSUES TO ADDRESS WHEN THE RESEARCH INVOLVES PROTECTED HEALTH INFORMATION

46. As part of this study, the researcher(s) will:
☐ Collect protected health information* from participants → Please complete question 47.
☐ Have access to protected health information* in the participants' records → Please complete question 47.
☒ None of the above → Please skip to question 48.

*Protected Health Information (PHI) is defined under HIPAA (Health Insurance Portability and Accountability Act of 1996) as health information transmitted or maintained in any form or medium that:

A. identifies or could be used to identify an individual;
B. is created or received by a healthcare provider, health plan, employer or healthcare clearinghouse; and
C. relates to the past, present or future physical or mental health or condition of an individual; the provision of health care to an individual; or the past, present or future payment for the provision of healthcare to an individual.

47. To use PHI in research you must have approval through one of the following methods:
A. An authorization signed by the research participant that meets HIPAA requirements; or
B. Use of a limited data set under a data use agreement.

Check below to indicate which method of approval you will use.

☐ A. Research participants in this study will sign an *Authorization to Use or Disclose PHI for Research Purposes* form. If the study includes multiple activities (e.g., clinical trial or collection and storage of PHI in a central repository), then two authorization forms must be submitted for review. You may download a sample authorization form at the IRB Web site, fill in the required information, and fax to (626) 605-0472.
☐ B. I will access a limited data set by signing a Data Use Agreement with the party that releases the PHI. A limited data set must have all possible identifiers removed from the data. It is the responsibility of the researcher and the party releasing the PHI to have in place and maintain a copy of a Data Use Agreement which meets HIPAA requirements. Use the template Data Use Agreement and fill in the required information. A copy of the signed Data Use Agreement must be submitted for IRB review.

POTENTIAL CONFLICTS OF INTEREST

48. This item asks you to disclose information relevant to separating your multiple roles as clearly as possible, with the goal of ensuring authentically <u>voluntary</u> participation in your study. Doctoral research directly benefits the student (allowing him or her to obtain a degree), and so the researcher should minimize the potential for either (a) conflict of interest or (b) perceived coercion to participate. Researchers who are in positions of authority must take extra precautions to ensure that potential participants are not pressured to take part in their study. <u>Data collection should be as detached as possible from the researcher's authority</u>.
Examples:
-a professor researcher may recruit students AFTER grades have been assigned
-a psychologist researcher may recruit clients from ANOTHER psychologist's practice
-a manager researcher may conduct ANONYMOUS data collection so that subordinates do not perceive their responses or [non]participation as being associated with their job standing

At the time of study recruitment, are the potential study participants aware of any of the researchers' other professional or public roles? (Such as teacher, business owner, community leader, supervisor, etc.?)
☒ No.
☐ Yes, at the time of recruitment some of the participants are aware of the researcher's <u>(Insert title)</u> role, and the following measures will be taken to separate the researcher's dual roles and minimize perceived coercion to participate: _____.

49. This item asks you to disclose information related to possible financial conflicts of interest, with the goal of maintaining research integrity. Is it possible that the financial situations or professional positions (to include promotions, contracts, clients, and reviews) of the researchers or their families could be directly impacted by the design, conduct, or results of this research?
☒ No.
☐ Yes, and the conflict of interest is being managed by the following disclosures/measures: _____.

50. Will the researcher give participants or stakeholders any gifts, payments, compensation, reimbursement, free services, or extra credit? It is acceptable to compensate your participants as long as the compensation cannot be interpreted as coercive among the participant population. For example, a $5 gift card to a coffee house is fine as a thank you gift, but an Ipod would not be, especially if the participants are teenagers. It is often better to eliminate compensation all together or make sure that 100% of your sample gets the same compensation (as opposed to only compensating those in your experimental group).

☒ No.
☐ Yes. More information is provided below.
    What compensation will be given? _____
    At what point during the research will the compensation be given? _____
    Under what conditions will the compensation be given? (i.e., how will compensation for withdrawn participants be handled?) _____

INFORMED CONSENT

This application's final checklist will direct you to email *unsigned drafts* of your consent/assent forms to IRB@waldenu.edu at the same time you submit this IRB form. Your application is not considered complete until they are received.

51. Federal regulations require that the informed consent procedures disclose each of the elements in the checklist below and that consent be documented (usually by asking the participants to sign the consent form listing all of the disclosures but there are some other arrangements that are acceptable, depending on the privacy issues and logistics of the data collection).

Anonymous surveys rely on implicit endorsement rather than obtaining a signed endorsement. In other words, instead of collecting a signature the researcher might instruct the participant to complete the survey if they agree to participate in the study as described in the cover page, which includes all the elements of informed consent below.

When participants are 6 and under, researchers must obtain parental consent in addition to reading a script that asks the children for their verbal assent to participate. When participants are between 7 and 17, researchers must obtain parental consent in addition to reviewing an age-appropriate assent form with the child and asking the child to sign if they want to participate. You may link to the relevant regulations from the Walden IRB Web site.

Templates for consent and assent forms can be downloaded from Walden IRB Web site. Note that the consent and assent forms on the IRB Web site are only templates and will likely need a great deal of tailoring for your study. Pay particular attention to making the reading level appropriate for your targeted participant population.

Please affirm that your consent/assent form(s) contain each of the following required elements.	YES	N/A
Statement that the study involves research	☒	
Statement of why subject was selected	☒	
Disclosure of the identity and all relevant roles of researcher (e.g., doctoral.student, part-time faculty member, facility owner)	☒	
An understandable explanation of research purpose	☒	
An understandable description of procedures	☒	
Expected duration of subject's participation	☒	
Statement that participation is voluntary	☒	
Statement that refusing or discontinuing participation involves no penalty	☒	
Description of reasonably foreseeable risks or discomforts	☒	
Description of anticipated benefits to subjects or others	☒	
Information on compensation for participation	☒	
Description of how confidentiality will be maintained	☒	
Whom to contact with questions about the research	☒	
Statement that subject may keep a copy of the informed consent form	☒	
All potential conflicts of interest are disclosed	☒	
Consent process and documentation are in language understandable to the participant	☒	
There is no language that asks the subject to waive his/her legal rights	☒	
If appropriate, indicates that a procedure is experimental (i.e., not a standard Rx)	☐	☒
If appropriate, disclosure of alternative procedures/treatment	☐	☒
If appropriate, additional costs to subject resulting from research participation	☐	☒

FINAL IRB CHECKLIST

52. Please indicate below which method you are using to send each of your supporting documents. We ask that you send these supporting documents to the IRB at the same time you submit this application.

Students must obtain their supervising faculty member's approval on the last page <u>before</u> submitting any materials to the IRB.

	Emailed to IRB@waldenu.edu	Faxed to (626) 605-0472	Not applicable to my study
Human Research Protections training completion certificate	☒	☐	
Data collection tools (e.g., surveys, interviews, assessments, etc.)	☒	☐	☐
All of the following that apply to any assessments' copyright holders: written/emailed permission to use the instrument, permission to reproduce the instrument in the dissertation, confirmation that the tool is public domain, proof of the researcher's qualifications to administer the instrument	☐	☐	☒
Letters of Cooperation from community partner organizations (e.g., school) or individuals (e.g. cooperating teacher) who are assisting with participant recruitment or data collection	☒	☐	☐
Data Use Agreement from any community partners that will be sharing their non-public records	☐	☐	☒
Invitation to participate in research (e.g., letter, flier, phone script, ad, etc.)	☒	☐	☐
Signed Confidentiality Agreements for transcribers, statisticians, research assistant, etc.	☐	☐	☐
Consent/assent forms	☒	☐	☐
Federal certificate of confidentiality (to shield data from subpoena)	☐	☐	☒

Please maintain a copy of this completed application for your records. Once the IRB application and all supporting documents have been received, the IRB staff will email the researcher and any relevant faculty supervisors to confirm that the IRB application is complete. At this time, the IRB staff will also notify the researcher of the expected IRB review date for the proposal.

The review date will be scheduled no later than 15 business days after your completion of this application. In the case of doctoral students, the review date will be scheduled no later than 15 business days after both A) the application is complete and B) the proposal is fully approved.

Notice of outcome of the IRB review will be emailed to the researcher and any supervising faculty members within 5 business days of the review. Please be aware that the IRB committee might require revisions or additions to your application before approval can be granted.

Neither pilot nor research data may be collected before notification of IRB approval. Students collecting data without approval risk expulsion and invalidation of data. The IRB will make every effort to help researchers move forward in a timely manner. Please contact IRB@waldenu.edu if you have any questions.

RESEARCHER ELECTRONIC SIGNATURE

53. By checking each of these boxes and providing my email address below as an authentication, I am providing an electronic signature certifying that each of the statements below is true.

☒ The information provided in this application form is correct, and was completed after reading all relevant instructions.
☒ I agree to conduct this and all future IRB correspondence electronically, via email/fax.
☒ I, the researcher, will request IRB approval before making any substantive modification to this study using the Request for Change in Procedures Form found at the Walden IRB Web site.
☒ I, the researcher, will report any unexpected or otherwise significant adverse events and general problems within one week using the Adverse Event Reporting Form found at the Walden IRB Web site.
☒ Neither recruitment nor data collection will be initiated until final IRB approval is received from IRB@waldenu.edu.
☒ I understand that this research, once approved, is subject to continuing review and approval by the Committee Chair and the IRB.
☒ I, the researcher, will maintain complete and accurate records of all research activities (including consent forms and collected data) and be prepared to submit them upon request to the IRB.
☒ I understand that if any of the conditions above are not met, this research could be suspended and/or not recognized by Walden University.

Researcher email address (provides authentication for electronic signature and thus must match email address on file with Walden University)	ml.dantzker@waldenu.edu

IRB Policy on Electronic Signatures

Walden's IRB operates in a nearly paperless environment, which requires reliance on verifiable electronic signatures. Electronic signatures are only appropriate when the signer is either (a) the sender of the email, or (b) copied on the email containing the signed document.

Electronic signatures are regulated by the Uniform Electronic Transactions Act. Legally, an "electronic signature" can be the person's typed name, their email address, or any other identifying marker. An electronic signature is just as valid as a written signature as long as both parties have agreed to conduct the transaction electronically. University staff will verify any electronic signatures that do not originate from a password-protected source (i.e., an email address officially on file with Walden).

SUPERVISING FACULTY MEMBER ELECTRONIC SIGNATURE

54. As the faculty member supervising this research, I assume responsibility for ensuring that the student complies with University and federal regulations regarding the use of human participants in research. By checking each of these boxes and providing my email address below as an authentication, I am providing an electronic signature certifying that each of the statements below is true.

☒ I affirm that the researcher has met all academic program requirements for review and approval of this research.
☒ I will ensure that the researcher properly requests any protocol changes using the Request for Change in Procedures Form found at the Walden IRB Web site.
☒ I will ensure that the student promptly reports any unexpected or otherwise significant adverse events and general problems within 1 week using the Adverse Event Reporting Form found at the Walden IRB Web site.
☒ I will report any noncompliance on the part of the researcher by emailing notification to IRB@waldenu.edu.

Faculty member email address (provides authentication for electronic signature and thus must match email address on file with Walden University):	smahony@waldenu.edu

IRB Policy on Electronic Signatures

Walden's IRB operates in a nearly paperless environment, which requires reliance on verifiable electronic signatures. Electronic signatures are only appropriate when the signer is either (a) the sender of the email, or (b) copied on the email containing the signed document.

Electronic signatures are regulated by the Uniform Electronic Transactions Act. Legally, an "electronic signature" can be the person's typed name, their email address, or any other identifying marker. An electronic signature is just as valid as a written signature as long as both parties have agreed to conduct the transaction electronically. University staff will verify any electronic signatures that do not originate from a password-protected source (i.e., an email address officially on file with Walden).

AUTHOR INDEX